T0320434

Risk and Uncertainty for Civil Engineering

This textbook introduces the fundamental concepts of probability, risk, and uncertainty, and shows their relevance in civil engineering projects. With an emphasis on applied probability and statistics, the book aids students in developing an intuitive understanding of the methods to apply in practice.

Drawing from real-world examples, readers are introduced to risk assessment and analysis techniques, enabling them to identify, evaluate, and prioritise potential risks. Examples are provided of how spreadsheet tools such as Microsoft Excel can be used to solve problems involving probability, and practical approaches such as Monte Carlo simulations and decision trees are explained in a clear, accessible manner, empowering students to make informed decisions under uncertain and variable conditions. The book also emphasises the importance of effective communication, equipping students with essential skills for working in multidisciplinary teams and with different stakeholders.

Risk and Uncertainty for Civil Engineering serves as an introductory textbook for undergraduate students in civil engineering, as well as a useful primer for postgraduate students.

Thomas Rodding Kjeldsen is Reader in the Department of Architecture and Civil Engineering, University of Bath, UK, where he teaches introductory courses on engineering risk and uncertainty to undergraduate and postgraduate civil engineering students.

Risk and Uncertainty for Civil Engineering

Thomas Rodding Kjeldsen

CRC Press is an imprint of the
Taylor & Francis Group, an **informa** business

Cover image: Svetlana Lukienko/Shutterstock

First edition published 2025
by CRC Press
4 Park Square, Milton Park, Abingdon, Oxon, OX14 4RN

and by CRC Press
2385 NW Executive Center Drive, Suite 320, Boca Raton FL 33431

CRC Press is an imprint of Informa UK Limited

British Library Cataloguing-in-Publication Data
A catalogue record for this book is available from the British Library

ISBN: 978-1-032-70032-8 (hbk)
ISBN: 978-1-032-67940-2 (pbk)
ISBN: 978-1-032-70037-3 (ebk)

DOI: 10.1201/9781032700373

Typeset in Sabon
by Apex CoVantage, LLC

Contents

Preface

INTENDED AUDIENCE

The material presented in this book has been developed by the author to introduce quantitative aspects of probability, risk, and uncertainty into the undergraduate and postgraduate civil engineering programmes at the University of Bath. The text assumes that readers have completed an introductory course in calculus and are familiar with concepts such as integration, differentiation, and Taylor expansions. There is also an implicit assumption that the audience is familiar with the use of EXCEL at a basic level.

The text is designed to help students become familiar with fundamental probability and statistical theory and methods and learn how these concepts can be used to quantify the risk and uncertainty in the engineering design process. Specifically, the learning objectives can be summarised as follows:

- Understand and apply basic probability theory for problems involving variability and uncertainty.
- Use basic statistical methods for exploratory data analysis, model identification, parameter estimation, and decision-making.
- Apply both analytical and numerical methods for describing the reliability of infrastructure with examples from geotechnical, structural, and hydraulic engineering.
- Gain knowledge of how best to communicate risk and uncertainty to both lay people and engineering experts.

The book provides examples of how spreadsheet tools such as EXCEL can be used to solve some problems involving probability. In my experience most undergraduate students are familiar with EXCEL. Other choices are available, including modern script-based programming environments such as R or Python. However, while some undergraduate civil engineering students undoubtedly master these programming languages to a high level, this is not the case for all. Thus, a spreadsheet approach was chosen to remove entry barriers to the material.

The emphasis of the material is applied probability and statistics, and the aim of the book is not to provide a comprehensive and rigorous introduction (including proofs) but rather to try and provide an intuitive understanding of the material, allowing students to apply the methods in practise. I have deliberately tried to focus all examples on situations of some relevance to civil engineering, opting to omit classical introductory examples, such as coin flips, dice rolls, and black and white balls, as much as possible.

Acknowledgements

The author is indebted to all students who have provided invaluable and constructive feedback on the material through the years. I am especially grateful to those who later wrote back and told me how much they enjoyed the course and how they have used the material in their professional careers to solve real-world engineering problems. I am grateful to Dr Mohammad Heidarzadeh and Emily Ingle for generously offering their time to read and comment on earlier versions of the current manuscript.

Chapter 1

Introduction

1.1 IMPORTANCE OF RISK AND UNCERTAINTY IN CIVIL ENGINEERING

Safety is a primary concern in all civil engineering design as failure of infrastructure can have severe social, economic, and environmental consequences. There is also a growing realisation that aspects of risk and uncertainty are a key part of all decision-making, including engineering design. This includes both technical assessments of risk and uncertainty such as designing to a particular probability of failure, but also more nebulous tasks such as communicating the existence of such uncertainties to the public at large. It is therefore important that professional civil engineers have, at least, some appreciation of these topics when they embark on their professional careers and in their continued practice.

1.2 DETERMINISTIC VERSUS STOCHASTIC WORLD VIEWS

Most traditional engineering calculations are based on a deterministic approach where all components of a system, as well as all external forces, are assumed perfectly known and predictable over the lifetime of the structure. In practise this is, at best, a working assumption. In reality the exact values of neither the load on the structure nor the resistance (i.e. ability of the structure to withstand the load) are known with absolute certainty. Consider for example the maximum wind-load on a building over its entire lifetime (e.g. 100 years). Due to the complexity of atmospheric processes it is not possible to predict the maximum wind speed over such an extended time period. In contrast, a risk-based approach to design starts from the premise that our knowledge of the load is uncertain and represented as a set of possible values, with each value assigned a probability (likelihood) elucidated from statistical analysis of past observations or from expert knowledge. Figure 1.1 illustrates the difference between a deterministic (solid arrow) and probabilistic representation of a design load. The x-axis represents all possible values of the

DOI: 10.1201/9781032700373-1

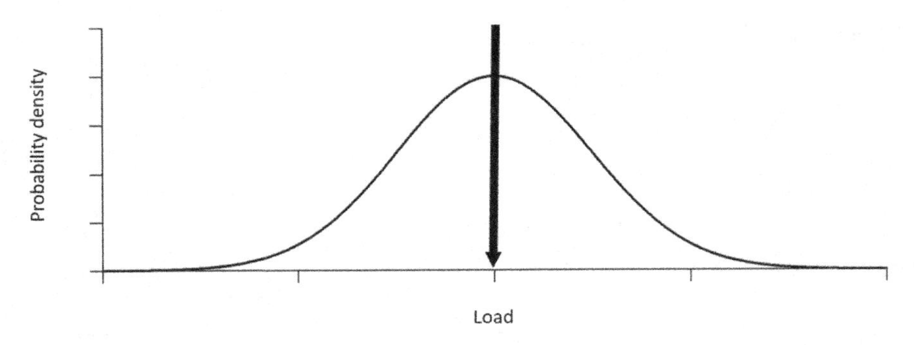

Figure 1.1 Deterministic (solid arrow) and stochastic (curve) representation of design loads.

design load and the y-axis the associated level of probability assigned to each value. In a deterministic world, the entire probability mass is consolidated in a single point (the arrow). The probabilistic representation (the curve) acknowledges that a range of possible future values are possible, but that some are more likely than others, including very extreme events.

Similarly, the resistance of the structure (defined here as the ability to withstand the external loading) is also more truthfully represented as a set of probabilities than a single deterministic value. For example, the strength of a material or soil foundation might not be known exactly due to variations in material production or heterogeneity of the soil matrix.

Many existing engineering design approaches acknowledge the inability to determine loads and resistance accurately in design via the introduction of concepts such as factor of safety and freeboards. However, such concepts still embrace a deterministic approach and are largely based on a rule-of-thumb approach to dealing with unknown aspects of the design process. It is in this context the terms "risk" and "uncertainty" should be understood. Uncertainty is the consequence of lack of knowledge, e.g. the exact strength of a concrete beam or the water demand of a city 20 years into the future.

1.3 DEFINITIONS OF RISK AND UNCERTAINTY

There are multiple definitions of terms such as risk and uncertainty. In engineering disciplines, risk is often defined as the product of the likelihood (or probability) and the associate consequence of a particular event. Consider an event A that has a probability of occurrence $P(A)$, and should A occur then the consequence is $C(A)$. In this case the risk of event A is defined as:

$$Risk(A)=P(A)C(A) \tag{1.1}$$

An example of A could be failure of a flood embankment, in which case $P(A)$ is the probability of the bank failure and $C(A)$ is the consequence of the failure measured in, for example, monetary damage. In this book the emphasis is on the calculation of the probability with less emphasis on the consequence aspect, as the latter is often very dependent on the exact circumstances of the problem.

It is more difficult to provide a unique definition of uncertainty, but the term implies imprecision, a lack of knowledge, ambiguity in answers, or an inability to provide an accurate forecast. The use of probability theory is an attempt to consider and quantify uncertainty, and students should master these technical aspects on completion of the book. The existence of uncertainty is uncomfortable, even if we can quantify it, as it makes decision-making and communication more difficult. This text acknowledges these practical implications of uncertainty by covering topics of decision-making, hypothesis testing, and communication.

1.4 THE ROLE OF PROBABILITY THEORY AND STATISTICS

Probability theory is the mathematical language allowing for a fundamental understanding of randomness and a quantitative assessment of the likelihood of future events. The theory of probability is a relatively recent innovation in mathematics. The modern version originated in the 17th century in an attempt to understand games involving gambling. Since then, the discipline has evolved and is now used to study random events across science, engineering, medicine, finance, and many, many more subjects. The parallel discipline of statistics has evolved alongside probability and is primarily concerned with extracting information from observed data and supporting decision-making under uncertainty. Probability and statistics are closely linked, and both are relevant to civil engineering, therefore aspects of both will be covered in this book.

Probability theory is often perceived as a difficult topic when first encountered despite the relative few fundamental rules governing probability. In the experience of authors this is, in part, because of difficulties embracing the ideas of random variables, i.e. bridging the gap between the deterministic and stochastic world views discussed in section 1.2. Thus, before discussing more advanced methods for analysing risk and uncertainty in the context of civil engineering, it is important to have a solid understanding of the basic rules. However, engineers are not mathematicians, so the emphasis of this book is on understanding and application, rather than fundamental theory.

1.5 BRIEF OUTLINE OF THIS BOOK

This book reflects what in the author's opinion can reasonably be fitted into a semester-long unit introducing risk and uncertainty in the second or third year of a civil engineering undergraduate programme. It should also be acknowledged that this is likely the first and last substantial text on probability, risk, and uncertainty many of these students will see. Thus the material has been selected and presented with the ambition of being accessible and of general interest, with enough of a steer for people who want to delve more into the details in further studies and application.

The complete outline of the book can be found in the table of contents, but what follows is a short commentary on the structure as a guide for students and instructors.

Chapters 2–4 cover an introduction to the basics of probability theory and statistical distributions. Chapter 2 introduces eight different distributions that are commonly used in civil engineering literature. Chapter 3 covers estimation theory, i.e. how to estimate distribution parameters to a sample of data using the method of moment and the maximum likelihood method, including a discussion of censored data. Both chapters include appendices summarising the main result of the eight chosen distributions. Chapter 4 introduces some standard plot types used for visualising data, including histograms, box plots, and frequency plots. In addition, the basics of how to create informative data visualisations are discussed, leaning on the work of Tufte (2001).

Chapters 5–7 take the basic concepts discussed in the early chapters and apply them to civil engineering systems, introducing methods for calculating and measuring the probability of failure. Chapter 5 focusses on approximate methods, starting from the mean and variance of linear systems and then moving on to the application of Taylor approximations to non-linear and multivariate systems. Chapter 6 discusses the use of computational Monte Carlo simulations to analyse failure probabilities of civil engineering systems. Finally, Chapter 7 covers limit state functions and the reliability index. The Hasofer-Lind method is introduced and accompanied by both analytical and EXCEL-based computation examples.

Chapters 8–13 focus on selected and more advanced topics. Complex systems are covered in Chapter 8 by describing the fundamental characteristics of series and parallel systems. The chapter concludes with an introduction to fault tree analysis drawing on the system concepts introduced earlier in the chapter. Chapter 9 is an introduction to decision-making under uncertainty, starting with a basic definition of a single decision-maker model including actions, states of nature, payoff, and opportunity loss. These concepts are used to introduce more advanced decision-making tools such as decision trees, expected value, and ultimately, utility theory and expected utility. Chapter 10 is the first sojourn into hypothesis testing. This is a difficult topic when first approached – in particular, the coming together of sample distributions, hypothesis testing, significance levels, and statistical power. Rather than providing a list of ready-made numerical recipes for a range of situations where a student might encounter hypothesis testing, the emphasis of the chapter is on understanding the fundamental concepts and ideas. The chapter is confined to examples based on a normal distribution with a known standard deviation, with reference to the more complex case of t-tests based on estimated standard deviations. Linear regression is introduced in Chapter 11, focussing on a simple univariate model estimated using least squares. The material sets out to demonstrate how the model parameters are estimated and how the R^2 summary statistic is calculated in line with results produced by fitting a linear trend line using EXCEL. Next, the significance of model parameters is discussed using hypothesis testing. Finally, prediction uncertainty of a regression model is covered, showing how multiple sources of uncertainty come together into a final prediction uncertainty. Chapter 12 is a discussion of high impact, low probability events focussing on the philosophical approaches to analysing these events, including record-breakers, a perfect storm, and a black swan event. Chapter 13 covers risk communication to a wider audience, including factors influencing how individuals perceive information provided in the language of probability risk and uncertainty. The chapter ends with a series of recommendations to help facilitate public communication. Chapter 14 is a reference chapter introducing selected EXCEL functions and demonstrating their usefulness to solve numerical probability problems, and general data summaries and queries. This chapter is meant as an appendix students can consult when solving numerical problems involving standard distributions.

The material in the book is supported by a number of EXCEL spreadsheets demonstrating solutions to example problems introduced in the book.

REFERENCE

Tufte, E. R. (2001). *The visual display of quantitative information*. Cheshire, CT: Graphics Press.

Chapter 2

Random Variables and Probability Distributions

2.1 DEFINITION OF PROBABILITY

Probability theory is concerned with assigning and estimating the probability of future random events. While probability theory cannot predict the exact outcome of the next event, for example the outcome of single coin flip, it can tell us something about the long-term behaviour of events such as a 50% chance of heads or tails if we consider the average result over multiple coin flips. The characterisation of such long-term behaviour is essential in engineering for planning and design of infrastructure systems, and for design of experiments.

In the context of probability theory, an experiment is a situation where a random phenomenon occurs, and the outcome is observed. The *sample space* Ω contains all possible outcomes (or elements) of the experiment. An event is defined as a subset of the sample space $A \in \Omega$. For example, a six-sided die has a sample space $\Omega = \{1,2,3,4,5,6\}$. Rolling the die and observing the outcome is an experiment, and an event can be, for example, an outcome where the die shows a number larger than 3, i.e. $A = \{4,5,6\}$.

A probability measure is defined on the sample space Ω as a function P which takes values between 0 and 1. In particular, the probability of all possible events in the sample space equals 1, i.e.:

$$P(\Omega) = 1 \tag{2.1}$$

Examples of sample spaces include:

- Toss of a fair coin: $\Omega = \{\text{Head}, \text{Tail}\}$.
- Roll of a die: $\Omega = \{1,2,3,4,5,6\}$.
- Annual rainfall in Oxford: $\Omega = \{\text{Rainfall} \geq 0\text{mm}\}$.

An event A can be assigned a probability $0 \leq P(A) \leq 1$. Often the event A is defined, and the task is to calculate $P(A)$.

Note that the notation for the probability measure varies between textbooks; sometimes it is denoted P as in these notes, whereas others prefer to use Pr or even $Prob$.

DOI: 10.1201/9781032700373-2

2.2 RULES FOR CALCULATING PROBABILITIES

In many practical problems it is important to understand how events relate to each other and the associated probability of two or more events occurring. For this purpose, it is necessary to introduce set theory, which is a mathematical discipline concerned with how objects relate to each other. Consider two events A and B that both belong to the same sample space Ω. Using set theory, it is possible to say something general about the relationships that might exist between the two events. The following definitions are important to analyse problems involving random events:

- *Union*: all elements in A or B, or both $(A \cup B)$.
- *Intersection*: all elements common to both A and B $(A \cap B)$.
- *Complementary event*: all elements not contained in A $\left(A^C\right)$.
- *Disjoint events*: events A and B have no elements in common, for example $A =$ "heads in first toss" and $B =$ "tail in first toss" are two disjoint events as they cannot both occur simultaneously. Disjoint events are sometime referred to as mutually exclusive events, emphasising that the two events cannot take place at the same time.
- *Independent events*: events A and B are considered independent if the occurrence of one event does not affect the probability of the other event occurring.

A Venn diagram is a useful approach to visualise the relationship between different events. The basic building block of a Venn diagram is a box representing the entire sample space Ω. Inside the box, circles (or any other shapes) demarcate a set of outcomes representing an event. If different shapes overlap, then different events share common elements; this overlapping area is known as the *intersection* of events and is represented by the mathematical symbol $\cap$. The entire area defined by several events is the *union* of events and is represented by the mathematical symbol $\cup$. Consider two events A and B; the first four definitions in the preceding list are illustrated using Venn diagrams in Figure 2.1.

If we denote the probability of observing an event in subset A and B by $\mathrm{P}(A) \geq 0$ and $\mathrm{P}(B) \geq 0$, then the following rules of calculating probabilities exist: the additive rule, complementary event, multiplicative rule, and the law of total probability.

2.2.1 Additive Rule

The additive rule represents the union $\cup$ of events. If a probability statement involves an "OR" statement – for example, *What is the probability of event A OR event B?* – the answer is:

$$P(A \cup B) = P(A) + P(B) - P(A \cap B) \tag{2.2}$$

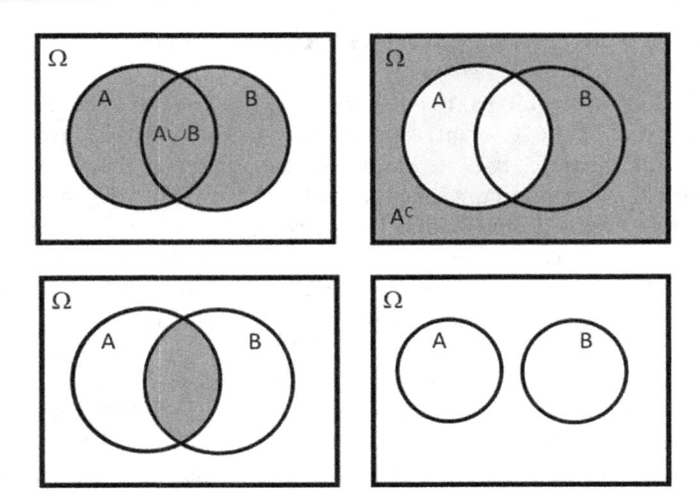

Figure 2.1 *Upper left:* Union of *A* and *B*. *Lower left:* Intersection of *A* and *B*. *Upper right:* Complementary event of *A*. *Lower right:* *A* and *B* are disjoint events.

If A and B are disjoint events, i.e. the occurrence of A precludes the occurrence of B (as per the definition above), then the probability of either A OR B is reduced to a simple addition:

$$P(A \cup B) = P(A) + P(B) \tag{2.3}$$

2.2.2 Complementary Event

The complementary event to event A is denoted A^C and considers all outcomes that are not included in event A. The probability of the complementary event A^C is:

$$P(A^C) = 1 - P(A) \tag{2.4}$$

Using the relationship in Eq. 2.4 is often useful, as for many situations it might be easier to calculate the probability of the complementary event rather than the event itself.

2.2.3 Multiplicative Rule

The multiplicative rule expresses the joint probability, or intersection, of two events A and B occurring simultaneously. If the probability statement involves an "AND" statement such as *What is the probability of A AND B occurring?*, then the answer is

$$P(A \cap B) = P(A|B)P(B) \tag{2.5}$$

Here, $P(A|B)$ is a conditional probability to be understood as the probability of A given that B has occurred (the vertical bar between A and B reads as "given that" or "conditional upon"). If A and B are independent events, then the probability of A given B is reduced to the probability of A, i.e. $P(A|B) = P(A)$, in which case Eq. 2.5 is reduced to simply:

$$P(A \cap B) = P(A)P(B) \tag{2.6}$$

For example, toss a coin twice and record the outcome. The outcome of the second coin toss does not depend on the outcome of the first coin toss; therefore the two events are independent. In contrast, the total volume of river runoff in one year depends on the total amount of rainfall that occurred in that same year; therefore runoff and rainfall cannot be considered independent events.

EXAMPLE 2.1 FUNDAMENTAL PROBABILITY CALCULATIONS

Consider a delivery of building elements to a construction site. According to the manufacturer of the elements, there is a probability of 0.05 that an element is faulty and should not be used. Two elements are randomly selected from the delivery and a set of different probability calculations undertaken. First, define the relevant events and their associated probability:

Event A: Element 1 is faulty, and $P(A) = 0.05$

Event B: Element 2 is faulty, and $P(B) = 0.05$

The complementary events to A and B are:

Event A^C: Element 1 is NOT faulty, $P(A^C) = 1 - P(A) = 1 - 0.05 = 0.95$

Event B^C: Element 1 is NOT faulty, $P(B^C) = 1 - P(B) = 1 - 0.05 = 0.95$

It is assumed here that the events are independent, i.e. that whether element 1 is faulty or not has no effect on the state of element 2. Also, the two events A and B are not mutually exclusive; it is possible for both element 1 and 2 to be faulty, or not faulty, at the same time.

What is the probability of element 1 <u>AND</u> 2 both being faulty?

$$P(A \cap B) = P(A)P(B) = 0.05 \times 0.05 = 0.0025$$

What is the probability of element 1 <u>OR</u> element 2 (or both) being faulty?

$$P(A \cup B) = P(A) + P(B) - P(A \cap B) = 0.05 + 0.05 - 0.05 \times 0.05 = 0.0975$$

What is the probability of element 1 being faulty AND element 2 being not faulty?

$$P(A \cap B^C) = P(A)P(B^C) = 0.05 \times 0.95 = 0.0475$$

What is the probability of one, but not both, elements being faulty?

The answer must consider the two possible outcomes: $(A \cap B^C)$ and $(A^C \cap B)$. It is important also to realise that these two events are mutually exclusive (disjoint events); i.e. they cannot both occur at the same time. Therefore, the answer is the probability that either of these events occur:

$$P((A \cap B^C) \cup (A^C \cap B)) = P(A \cap B^C) + P(A^C \cap B) = P(A)P(B^C) + P(A^C)P(B)$$
$$= 0.05 \times 0.95 + 0.95 \times 0.05 = 0.095$$

2.2.4 Law of Total Probability

The law of total probability allows for the probability of an event A to be expressed through a series of conditional probabilities. Let $B_1, \ldots, B_n$ be a set of n disjoint events $(B_i \cap B_j = 0 \text{ for } i \neq j)$ where $P(B_i) \geq 0$ and $\sum P(B_i) = 1$; then for an event A the law of total probability states that:

$$P(A) = \sum_{i=1}^{n} P(A|B_i)P(B_i) \tag{2.7}$$

The law of total probability is useful as sometimes it is easier to specify $P(A|B_i)$ and $P(B_i)$ than it is to specify $P(A)$ directly. An example of application of the law of total probability to a dam safety study is provided in Example 2.4.

2.3 RANDOM VARIABLES

This section introduces the concept of a random variable, which is key for being able to consider more complex calculations of risk and uncertainty.

A random variable is a function from the sample space Ω to the real numbers. As the events in Ω are characterised by being random, the random variable is characterised by a probability distribution. A more heuristic description might say that a random variable is a mathematical concept used to describe and quantify uncertain or random phenomena. Random variables are typically denoted by upper-case italic letters such as X, Y, and Z.

It is common to distinguish between two types of random variables: discrete and continuous random variables:

Discrete random variables can take on only countable numbers such as $0,1,2,3,\ldots$

- X: number of tails obtained in ten tosses with a fair coin. Possible values $X = 0,1,2,3\ldots,10$.
- X: number of planes arriving in Heathrow Airport in an hour. Possible values $X = 0,1,2,\ldots$

Continuous random variables can take on a continuum of values such as any value in the interval $-\infty$ to ∞, or from 0 to ∞. These types of variables are often used to represent phenomena that can be measured (as opposed to counted).

- X: the annual rainfall in Oxford. Possible values include all values of $X \geq 0\,\text{mm}$.
- X: waiting time until next bus arrives at university campus. Possible values include all values of $X \geq 0$ seconds.

2.4 PROBABILITY DISTRIBUTIONS

A continuous random variable is characterised by a particular probability distribution, which is a mathematical model representing the long-term probabilistic behaviour of the phenomenon described by the random variable. Loosely speaking, the probability distribution assigns a likelihood to each member of the subset Ω (i.e. all possible values). Many different probability distributions exist, and the choice depends on the type of problem and the data being analysed. However, common to all probability distributions is that they are characterised by a probability density function (pdf), which assigns likelihood to a particular outcome of a random variable, which, in turn, can be interpreted as a probability. For example, the probability that a random variable X has a value of exactly x is:

$$P(X = x) = f(x)dx \tag{2.8}$$

where the function $f(x)$ is the probability density function. Note here that the upper-case X is the random variable while the lower-case x is a specific numeric value.

In accordance with the definition of the sample space Ω in Eq. 2.1, integrating a pdf over the entire sample space should equal 1 (remember the probability of all possible events in the sampling space equals 1), i.e.

$$P(\Omega) = \int_{-\infty}^{\infty} f(x)dx = 1 \tag{2.9}$$

Starting from the pdf, the very important concept of the cumulative distribution function (cdf) is derived, which defines the probability of a random variable X taking on a value equal to or less than a certain value x and can be defined as:

$$F(x) = P(X \leq x) = \int_{-\infty}^{x} f(t)dt \tag{2.10}$$

where t is a dummy variable that is integrated out. Again, consider a specific value x. If an event A is defined as $X \leq x$, then the complementary event A^c is defined as $X > x$, and according to Eq. 2.4:

$$P(X \leq x) = 1 - P(X > x) \tag{2.11}$$

Graphical representations of four different probability statements in Eqs. 2.8, 2.9, 2.10, and 2.11 are shown in Figure 2.2 considering a standardised normal distribution (to be discussed in more detail in section 2.6.1). In addition, examples of the pdf and how to calculate the corresponding cdf for a continuous random variable are shown in Examples 2.2 and 2.3. All four figures show that the actual probability should be interpreted as the area under the pdf function rather than the value of the pdf itself.

A discrete random variable is characterised by a probability mass function (pmf) describing the probability that a random variable takes on a particular value x as:

$$f(x) = P(X = x) \tag{2.12}$$

The sum of the pmf over all possible values in the sample space Ω equals 1 as per Eq. 2.1, i.e.

$$\sum_{x \in \Omega} f(x) = 1 \tag{2.13}$$

where $x \in \Omega$ signals that the sum is considered over all values of x that are members of Ω. The cumulative distribution function (cdf) for a discrete random variable is the sum of all probabilities assigned to elements equal to or lower than the value x, i.e.

$$F(x) = \sum_{t \leq x} f(t) \tag{2.14}$$

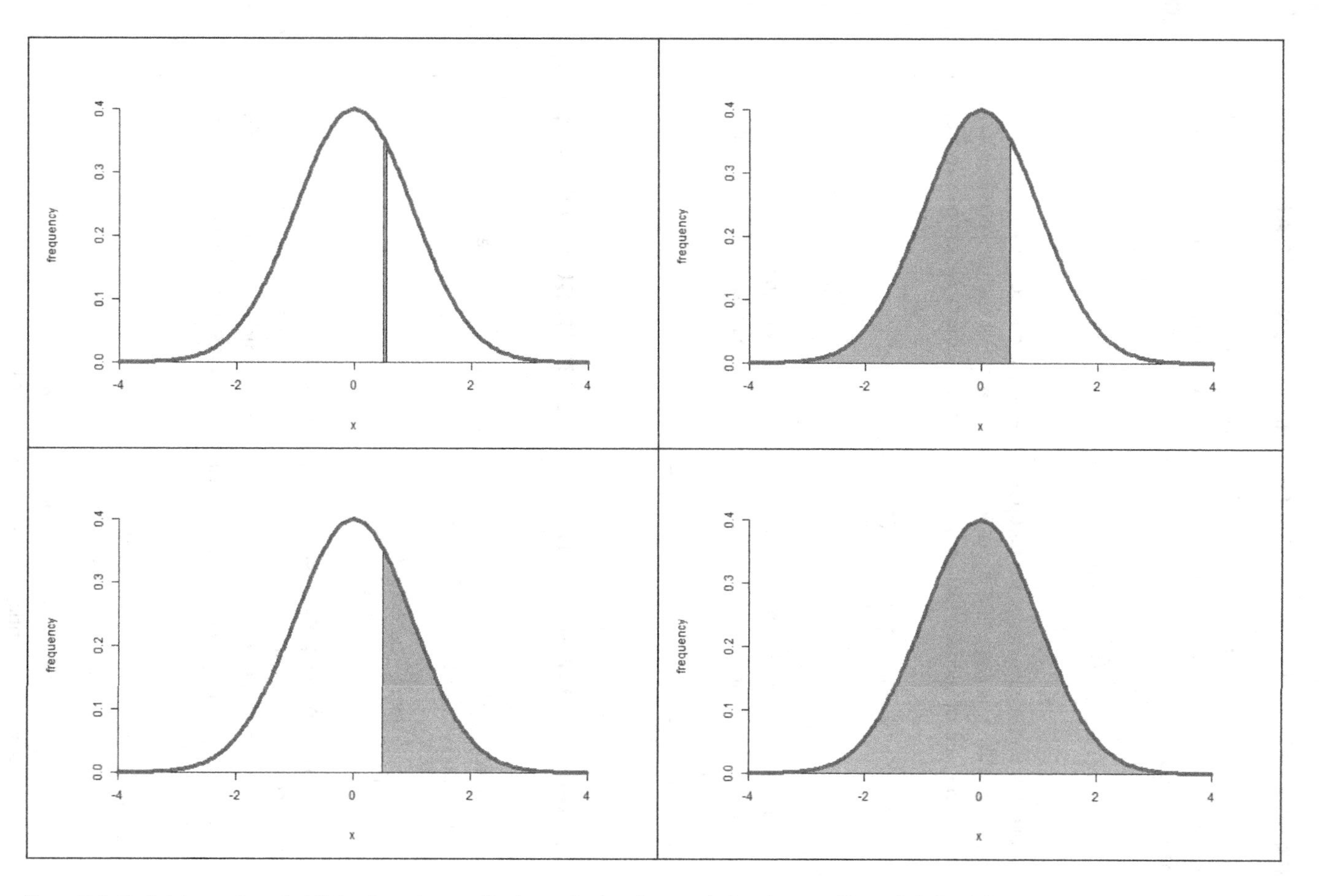

Figure 2.2 Definitions of probabilities based on pdf of standardised normal distribution. *Upper left*: Eq. 2.8. *Upper right*: Eq. 2.10. *Lower left*: Eq. 2.11. *Lower right*: Eq. 2.9.

EXAMPLE 2.2 THE POISSON DISTRIBUTION, PMF, AND CDF

A discrete random variable X can take all positive integer values (including zero). Consider the random variable to follow a Poisson distribution with a pmf defined as:

$$f(k) = P(X = k) = \frac{\lambda^k}{k!} exp(-\lambda),\ k = 0, 1, 2, \cdots$$

where λ is a model parameter. The upper-case X is the random variable while the lower-case k is a specific numerical value. The cdf of the discrete distribution is the sum of all individual contributions to the cumulative probability, i.e.

$$F(k) = P(X \le k) = \sum_{i=0}^{k} f(i) = \sum_{i=0}^{k} \frac{\lambda^i}{i!} exp(-\lambda)$$

where $f(i)$ is the pmf evaluated for the value i.

The Poisson distribution can describe the number of occurrences in a given time interval, i.e. number of earthquakes per year or number of cars passing a location per minute.

EXAMPLE 2.3 THE EXPONENTIAL DISTRIBUTION, PDF, AND CDF

A particularly simple continuous distribution is the one-parameter exponential distribution, which has a pdf defined as

$$f(x) = \begin{cases} \lambda \exp(-\lambda x) & x \ge 0 \\ 0 & x < 0 \end{cases}$$

where λ is a model parameter. The exponential distribution is often used for describing the random nature of waiting times, such as the waiting time between bus arrivals or the waiting time between major flood or earthquake events. The cdf for the exponential distribution is obtained by integrating the pdf as per Eq. 2.10, i.e.

$$F(x) = \begin{cases} 1 - \exp(-\lambda x) & x \ge 0 \\ 0 & x < 0 \end{cases}$$

The two graphs below show the relationship between the pdf (top) and the cdf (bottom) for an exponential distribution with $\lambda = 1$.

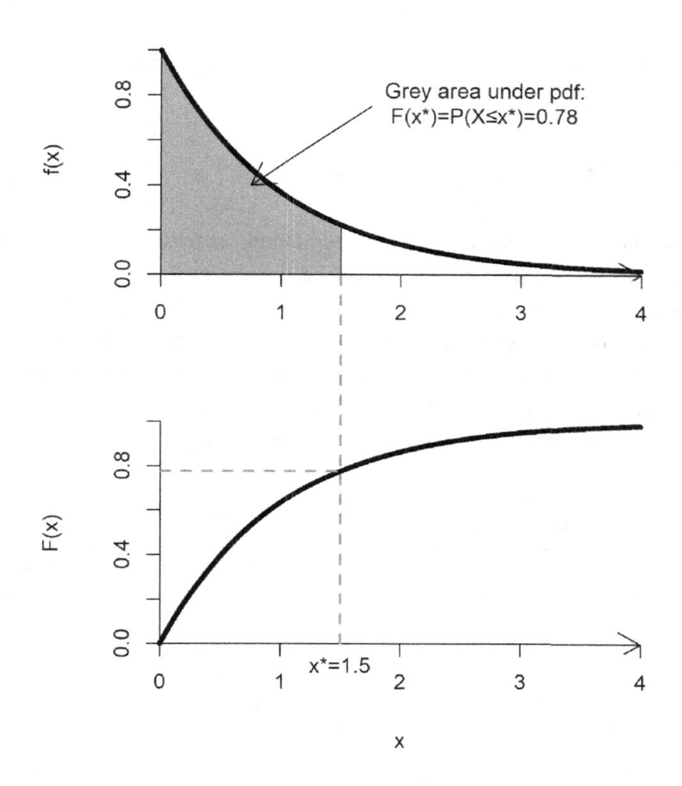

For $x^* = 1.5$, the probability that $X \leq x^*$ is represented by the grey area on the top figure (the area under the pdf) and is calculated by evaluating the cdf at x^*, which for $\lambda = 1$ gives

$$P\left(X \leq x^*\right) = F\left(x^*\right) = 1 - \exp\left(-x^*\right) = 0.78$$

Thus the values of the cdf on the lower graph correspond to the corresponding area (grey area) under the pdf (top graph).

EXAMPLE 2.4 PROBABILITY OF RESERVOIR FAILURE

An earth embankment dam has a primary free overflow spillway and a secondary spillway with two identical mechanically operated gates (Gate 1 and Gate 2) that can be opened in a situation where additional spillway capacity is needed.

Assume that the free spillway has a capacity of 100 m^3/s and never fails. Each of the two gates on the second spillway has a capacity to release 250 m^3/s and operate fully independent of the other. The peak of the inflow hydrograph,

Q (m^3/s) to the dam in the case of a flood event is distributed as a Gumbel distribution with a cdf given as:

$$F(q) = P(Q \leq q) = exp\left(-exp\left(-\frac{q-116}{58}\right)\right)$$

The dam fails if the inflow exceeds the combined capacity of the two spillways (primary and secondary), in which case the uncontrollable spill over the dam crest will cause erosion of the embankment. To calculate the probability of failure of the dam resulting from overtopping (i.e. inadequate spillway capacity), it is necessary to consider first all the possible failure conditions. As shown in the table below, there are four different scenarios where combinations of inflow and available spillway capacity will result in overtopping.

Four Mutually Exclusive Scenarios Where Combinations of Gate Operation and Inflow Will Result in Dam Failure

Scenario	*Inflow Q (m^3/s)*	*Gate 1*	*Gate 2*
1	Q > 350	Fail	Operate
2	Q > 350	Operate	Fail
3	Q > 600	Operate	Operate
4	Q > 100	Fail	Fail

For example, the dam will fail if both Gate 1 and Gate 2 fail and if the inflow exceeds 100 m^3/s, which is the capacity of the free spillway. Similarly, if both the free spillway and one of the two gates operate, then the dam will fail if the inflow exceeds the combined capacity of 100 m^3/s + 250 m^3/s = 350 m^3/s. Finally, if both gates operate, then the dam will fail if the inflow exceed the total capacity of 100 m^3/s + 250 m^3/s + 250 m^3/s = 600 m^3/s.

Next, define two Events A and B as:

- Event A: Gate 1 is operational, and P(A) = 0.99.
- Event B: Gate 2 is operational, and P(B) = 0.95.

where the complementary events (gates fail) are denoted A^C and B^C, respectively. As expected, the probability of the gates being operational is relatively high to ensure that the dam is mostly operating as designed. The four event sets representing the conditions of the two gates listed in the first table can be written as shown in the new table below.

Event Sets Representing All Possible Combinations of Gate Failure or Operation

Gate 1	*Gate 2*	*Event Set*	*Probability*
Fail	Operate	$A^C \cap B$	$(1-0.99)\times 0.95 = 0.0095$
Operate	Fail	$A \cap B^C$	$0.99 \times (1-0.95) = 0.0495$
Operate	Operate	$A \cap B$	$0.99 \times 0.95 = 0.9405$
Fail	Fail	$A^C \cap B^C$	$(1-0.99)\times(1-0.95) = 0.0005$

These four events are mutually exclusive, meaning they cannot take place at the same time, i.e. a gate cannot be both operational and non-operational at the same time.

It is possible to draw the Venn diagram for combinations of these events. For example, the Venn diagram for the union of Scenario 1 and 2 from the first table (Gate 1 fails AND Gate 2 operates) OR (Gate 1 operates AND Gate 2 fails) is defined as

$$\left(A^C \cap B\right) \cup \left(A \cap B^C\right)$$

and the Venn diagram is shown in the figure below:

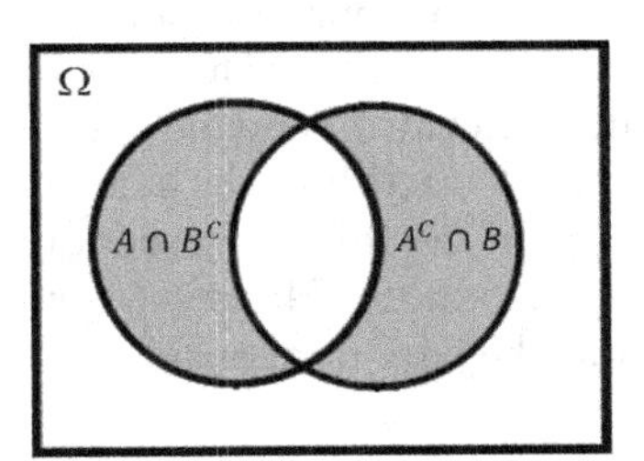

Venn diagram where the grey area represents $\left(A^C \cap B\right) \cup \left(A \cap B^C\right)$.

Finally, the probability of the dam failure can now be calculated based on the law of total probability. This is possible by considering the four failure scenarios conditional on the inflow exceeding capacity, while also recognising that the four scenarios are mutually exclusive. The law of total probability (Eq. 2.7) states that the total probability of failure is a combination of the probability of each of the four scenarios conditional on the inflow exceeding the capacity of the spillway in each scenario. So with reference to the event-set notation in the first table, the total probability of failure can be written as

$$P(\text{failure}) = P(A \cap B \mid Q > 600) \times P(Q > 600) +$$
$$P\left((A^c \cap B) \cup (A \cap B^c) \mid Q > 350\right) \times P(Q > 350) +$$
$$P(A^c \cap B^c \mid Q > 100) \times P(Q > 100)$$

The probability distribution representing inflow magnitude gives $P(Q > q) = 1 - F(q)$, where $F(q)$ is the Gumbel distribution:

$$P(\text{failure}) = 0.9405 \times (1 - F(600)) + (0.0095 + 0.0495) \times (1 - F(350)) + 0.0005 \times (1 - F(100)) = 0.00168$$

2.5 STATISTICAL MOMENTS

The distribution function $f(x)$ provides a complete description of the random variable X, but often it is useful to summarise the distribution using only one or two numbers. For this purpose, statistical moments are used to characterise the shape of distributions. The first statistical moment is the mean value and represents the central tendency of the data, while the second moment is the variance and characterises the general spread of the probability across the sample space. In the following sections the first two moments (mean and variance) are introduced. Higher-order moments exist, such as skewness and kurtosis, but they are not considered further here.

2.5.1 The Mean Value

The first statistical moment is known as the mean value and is defined through the mean-operator $\mu = E(X)$. For a discrete random variable the mean value is defined mathematically as:

$$\mu = E(X) = \sum_{i=0}^{\infty} x_i P(X = x_i) \tag{2.15}$$

while for a continuous random variable the equivalent definition is

$$\mu = E(X) = \int_{-\infty}^{\infty} x f(x) dx \tag{2.16}$$

The mean value is essentially a probability weighted average over all possible values of the random variable.

EXAMPLE 2.5 THE MEAN VALUE OF THE EXPONENTIAL DISTRIBUTION

Derive the mean value of the exponential distribution:

$$f(x) = \begin{cases} \lambda \exp(-\lambda x) & x \geq 0 \\ 0 & x < 0 \end{cases}$$

The exponential distribution describes a continuous random variable, and therefore the mean value is defined according to Eq. 2.15:

$$E(X) = \int_{-\infty}^{\infty} x f(x) dx = \int_{0}^{\infty} x\lambda \exp(-\lambda x) dx = \frac{1}{\lambda}$$

Note that the mean value of the distribution is expressed through the value of the model parameter λ.

2.5.2 Mean of a Function of a Random Variable

Sometimes it is not the mean of a random variable X itself but rather the mean of a function of a random variable $g(X)$ that needs to be evaluated. Fortunately, the definition of the mean of a function $Y = g(X)$ is defined for a discrete and continuous random variable as

$$E(Y) = \sum_{x \in \Omega} g(x) p(x) \tag{2.17}$$

$$E(Y) = \int_{-\infty}^{\infty} g(x) f(x) dx \tag{2.18}$$

where $p(x)$ and $f(x)$ are the probability mass function (pmf) and probability density function (pdf) describing a discrete and continuous random variable, respectively.

2.5.3 Variance and Standard Deviation

The variance of a random variable is a measure of the spread of the distribution around the mean value μ. The larger the spread, the larger the variance. Mathematically, the variance operator $\sigma^2 = V(X)$ is defined for a discrete random variable as

$$\sigma^2 = V(X) = \sum_{i=0}^{\infty} (x_i - \mu)^2 P(X = x_i) \tag{2.19}$$

and the equivalent definition for a continuous random variable is

$$\sigma^2 = V(X) = \int_{-\infty}^{\infty} (x-\mu)^2 f(x)dx \tag{2.20}$$

The standard deviation is represented by the Greek letter σ and is the square root of the variance. The units of the standard deviation are the same as the units of the mean value.

The coefficient of variation CV is defined as the ratio between the standard deviation and the mean:

$$CV = \frac{\sqrt{V(X)}}{E(X)} = \frac{\sigma}{\mu} \tag{2.21}$$

The coefficient of variation is also sometimes known as the relative standard deviation and expressed as a percentage.

EXAMPLE 2.6 THE VARIANCE AND COEFFICIENT OF VARIATION OF THE EXPONENTIAL DISTRIBUTION

Derive the variance of the exponential distribution:

$$f(x) = \begin{cases} \lambda \exp(-\lambda x) & x \geq 0 \\ 0 & x < 0 \end{cases}$$

The exponential distribution describes a continuous random variable, and therefore the variance is defined according to Eq. 2.20:

$$V(X) = \int_{-\infty}^{\infty} (x-\mu)^2 f(x)dx = \int_{0}^{\infty} (x-\mu)^2 \lambda \exp(-\lambda x)dx = \frac{1}{\lambda^2}$$

The coefficient of variation for the exponential distribution can be found using Eq. 2.21 by combining the result above with the expression of the mean value found in Example 2.5.

$$CV = \frac{\sqrt{V(X)}}{E(X)} = \frac{\sqrt{1/\lambda^2}}{1/\lambda} = 1$$

2.5.4 Covariance and Correlation

The covariance between two random variables X and Y is defined as:

$$Cov(X,Y) = E((X-\mu_X)(Y-\mu_Y)) \tag{2.22}$$

The subscripts have been added to emphasise that μ_X and μ_Y refer to the mean values of X and Y, respectively. Next, the correlation coefficient is defined as

$$\rho = \frac{Cov(X,Y)}{\sigma_X \sigma_Y} \tag{2.23}$$

Again, subscripts are added to emphasise that σ_X and σ_Y are the standard deviations of X and Y. The correlation coefficient can take on values between -1 and 1. Data are positively correlated when large values of X generally result in large values of Y, and negatively correlated if large values of X generally result in small values of Y.

There are some general rules for how to manipulate covariance terms. If X, Y, and Z are three random variables and a_0, a_1, a_2, and a_3 are constants, then:

$$Cov(a_0 + a_1X, a_2 + a_3Y + a_4Z) = a_1a_3Cov(X,Y) + a_1a_4Cov(X,Z) \tag{2.24}$$

Note that if X and Y are independent random variables, then $Cov(X,Y) = 0$, but that the reverse is not true, i.e. $Cov(X,Y) = 0$ does not necessarily mean that X and Y are independent. Finally, note that:

$$Cov(X,Y) = Cov(Y,X) \tag{2.25}$$

and

$$Cov(a_o X, a_1 X) = a_0a_1 V(X) \tag{2.26}$$

Sometimes in the engineering literature the coefficient of variation, denoted CV in Eq. 2.21, is defined as COV. It is important to understand the difference between CV and Cov as the terms are used in this note.

2.5.5 Quantiles of Distributions

Once a distribution has been specified for a random variable, it is possible to extract a value of the random variable associated with a specified non-exceedance probability. These values are denoted quantiles. For example, the 0.90 quantile is the value of a random variable not exceeded with a probability of 0.90.

Technically, quantiles are derived by inversing the cumulative distribution function (cdf). Consider the statement below, where a cdf evaluated in a point x_p is equal to a non-exceedance probability p as:

$$P(X \le x_p) = F(x_p) = p \tag{2.27}$$

Assume the probability p is known, then the pth quantile is the corresponding value of x_p which can be found by inversing the cdf as:

$$x_p = F^{-1}(p) \tag{2.28}$$

A commonly used summary statistic is the median, defined as the 0.50 quantile, i.e. the point on the distribution where 50% of the probability mass is below and the other 50% above.

EXAMPLE 2.7 THE MEDIAN (0.50 QUANTILE) OF THE EXPONENTIAL DISTRIBUTION

Derive the median of the exponential distribution with a cdf from Example 2.3. For all positive values of x the cdf is defined as:

$$F(x) = 1 - \exp(-\lambda x)$$

Using Eq. 2.28, the median can be found as the 0.50 quantile, i.e.:

$$x_{0.50} = F^{-1}(0.50)$$

Thus

$$1 - \exp(-\lambda x_{0.50}) = 0.50$$

$$x_{0.50} = \frac{\ln 2}{\lambda}$$

Note here that the median is different from the mean value derived in Example 2.5. While the mean value and median can both provide valuable information on the location of the distribution, they represent two different methods for doing so.

2.6 EXAMPLES OF PROBABILITY DISTRIBUTIONS

This section will introduce eight probability distributions that are all commonly used for risk calculations in civil engineering. However, these distributions are only a very small subsection of all the distributions that can be found in the more specialised statistical and technical literature.

A general notation for specifying that a random variable X follows a particular probability distribution *dist* with model parameters a and b is given as:

$$X \sim dist(a, b)$$

where the *dist* and the parameters are specified for each distribution. Methods for choosing distributions and estimating the parameter values based on data analysis are discussed in more details in Chapters 3 and 4.

2.6.1 The Uniform Distribution

As we shall see later, the uniform distribution plays an important part in advanced simulation studies. The distribution is characterised by a constant value of the pdf for all possible values of a random variable X defined on an interval with an upper and lower limit $X \sim U(a;b)$, and the pdf is:

$$f(x) = \begin{cases} \dfrac{1}{b-a} & a \leq x \leq b \\ 0 & \text{elsewhere} \end{cases} \tag{2.29}$$

The corresponding cdf is defined as

$$F(x) = \begin{cases} \dfrac{x-a}{b-a} & a \leq x \leq b \\ 0 & \text{elsewhere} \end{cases} \tag{2.30}$$

A graphical representation of the pdf and cdf of the uniform distribution is shown in Figure 2.3.

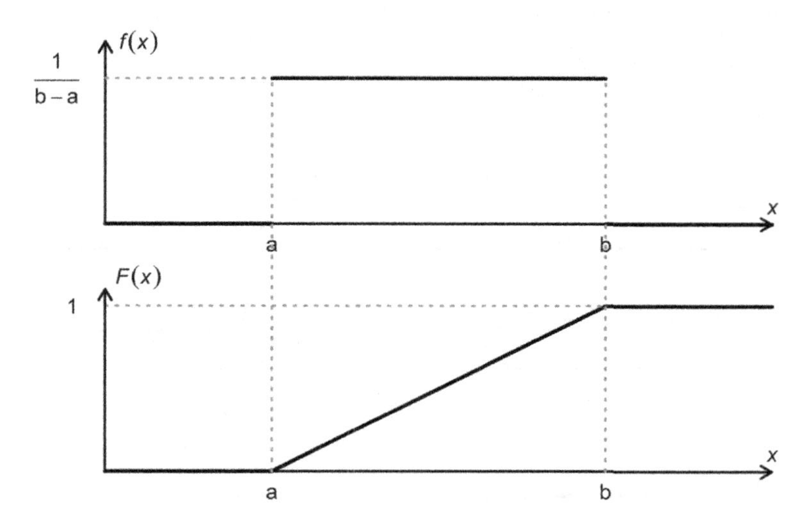

Figure 2.3 The pdf (upper) and cdf (lower) of uniform distribution between lower *a* and upper *b* limits.

The mean and variance of a random variable following a uniform distribution are given as follows:

$$E(X) = \frac{a+b}{2} \tag{2.31}$$

$$V(X) = \frac{(b-a)^2}{12} \tag{2.32}$$

2.6.2 The Normal Distribution

The normal distribution is the most important tool in engineering risk and uncertainty assessments. If a random variable X follows a normal distribution with parameters μ and σ^2, then $X \sim N(\mu, \sigma^2)$, and the pdf is given as:

$$f(x) = \frac{1}{\sqrt{2\pi}}\frac{1}{\sigma}\exp\left[-\frac{1}{2}\left(\frac{x-\mu}{\sigma}\right)^2\right], -\infty < x < \infty \tag{2.33}$$

The two parameters μ and σ^2 represent the mean value and variance. As this pdf cannot be integrated analytically, the cdf is defined as a definite integral of the pdf which must be solved numerically.

$$F(x) = \int_{-\infty}^{x} f(t)\,dt \tag{2.34}$$

A graphical representation of the pdf and cdf of the normal distribution is shown in Figure 2.4.

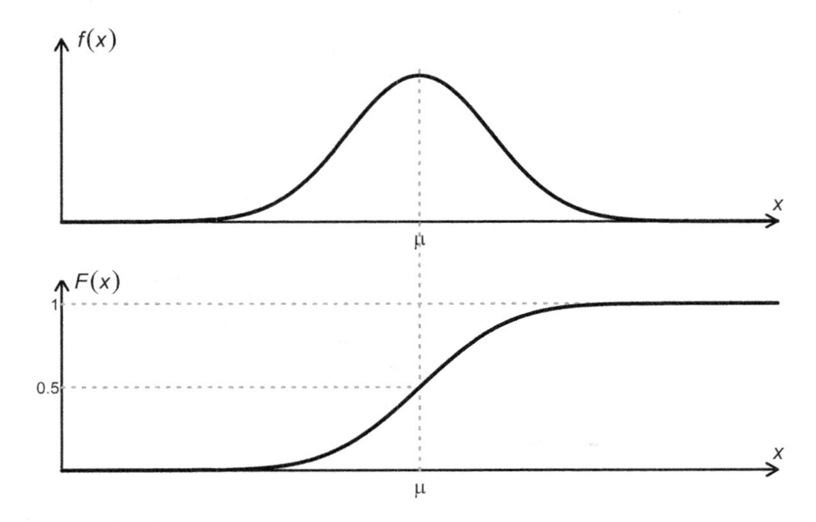

Figure 2.4 The pdf (upper) and cdf (lower) of normal distribution with mean value μ.

A special case of the normal distribution is the *standard normal distribution* Z, defined as having zero mean value and a variance of 1, i.e. $Z \sim N(0,1)$, where the pdf and cdf are defined using the lower-case and upper-case Greek letter phi, ϕ and Φ, as:

$$\phi(z) = \frac{1}{\sqrt{2\pi}} \exp\left[-\frac{1}{2}z^2\right] \tag{2.35}$$

$$\Phi(z) = \int_{-\infty}^{z} \phi(t)dt \tag{2.36}$$

As before, there is no explicit analytical solution to the cdf in Eq. 2.36. But values of $\Phi(z)$ can be found using statistical tables (see Appendix A in this chapter) or numerical software such as EXCEL R or Python.

The mean and variance of a random variable X following a normal distribution are given as:

$$E(X) = \mu \tag{2.37}$$

$$V(X) = \sigma^2 \tag{2.38}$$

If a random variable is normally distributed as $X \sim N(\mu, \sigma^2)$, then it can be defined in terms of the standard normal distribution as:

$$X = \mu + \sigma Z \Leftrightarrow Z = \frac{X - \mu}{\sigma} \tag{2.39}$$

where again Z is the standard normal distribution obtained in Eq. 2.39 by subtracting the mean and standardising with the standard deviation. This means that for any random variable X that is normally distributed, the following transformation can be made to facilitate calculations of probabilities:

$$P(X \leq x) = P\left(\frac{X - \mu}{\sigma} \leq \frac{x - \mu}{\sigma}\right) = P\left(Z \leq \frac{x - \mu}{\sigma}\right) = \Phi\left(\frac{x - \mu}{\sigma}\right) \tag{2.40}$$

Finally, the quantile of the distribution is defined as the value of x_p that corresponds to a fixed probability p in Eq. 2.40, i.e.:

$$\Phi\left(\frac{x_p - \mu}{\sigma}\right) = p \tag{2.41}$$

$$\Phi^{-1}(p) = \frac{x_p - \mu}{\sigma} \tag{2.42}$$

EXAMPLE 2.8 DETERMINE THE QUANTILE OF A NORMAL DISTRIBUTION

Find the value of $\Phi(z)$ for $z = 0.75$. The answer can be found manually by looking up the value in the table in Appendix A. $\Phi(0.75) = 0.77337$.

Alternatively, the EXCEL command: *NORMDIST* can be used as shown in Figure A:

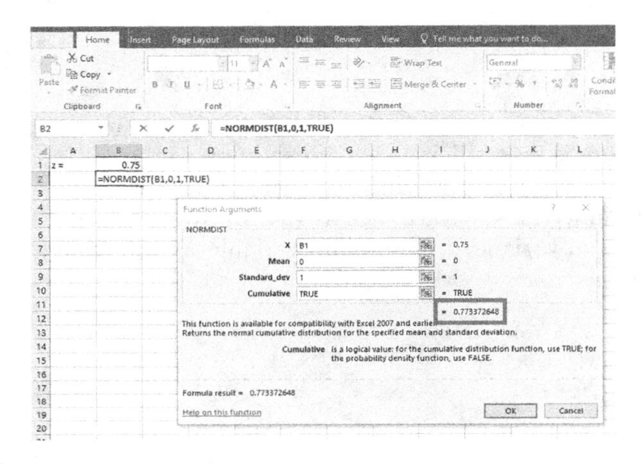

Figure A EXCEL *NORMDIST* function

Solving the inverse problem and determine the $p = 0.77337$ quantile in the standard normal distribution can be done using either the table in Appendix A or through the *NORM.INV* function in EXCEL as shown in Figure B:

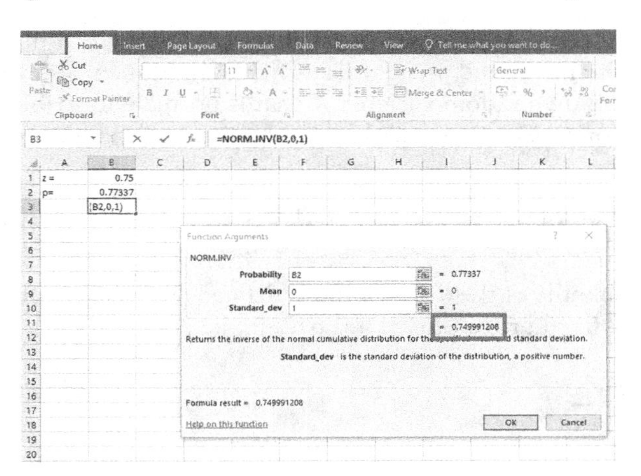

Figure B EXCEL *NORM.INV* function

Next, consider two independent random variables, both following a normal distribution $X \sim N(\mu_x, \sigma_x^2)$ and $Y \sim N(\mu_y, \sigma_y^2)$. The sum of X and Y is itself a normally distributed random variable such that:

$$X + Y \sim N(\mu_x + \mu_y, \sigma_x^2 + \sigma_y^2) \tag{2.43}$$

More generally, a linear combination of independent normally distributed random variables is itself normally distributed. Consider $i = 1, \ldots n$ normally distributed independent random variables $X_i \sim (\mu_i, \sigma_i^2)$, then the sum of these variables is normally distributed as:

$$a_0 + a_1 X_1 + \ldots + a_n X_n \sim N(a_0 + \sum a_i \mu_i, \sum a_i^2 \sigma_i^2) \tag{2.44}$$

where a_i, $i = 0, 1, \ldots, n$ is a series of constants (i.e. not random variables). In Chapter 5, an extension to this rule will be introduced considering correlated random variables.

EXAMPLE 2.9 SAFETY MARGIN

Consider a structural column where an external vertical load is a normally distributed random variable $Q \sim N(\mu_Q, \sigma_Q^2)$. The internal resistance of the column is denoted R and is also a normally distributed random variable $R \sim N(\mu_R, \sigma_R^2)$.

We define a new random variable Y as the difference between load and resistance, $Y = Q - R$. The structure is considered safe as long as the load is less than the resistance, i.e. $Y = Q - R < 0$. As both Q and R are normally distributed and independent, the distribution of difference $Y = Q - R$ is also a normal distribution according to Eq. 2.43, i.e.:

$$Y = Q - R \sim N(\mu_Q - \mu_R, \sigma_Q^2 + \sigma_R^2)$$

The probability of the column failing is therefore:

$$P(Y > 0) = 1 - P(Y \leq 0) = 1 - \Phi\left(\frac{0 - \mu_Q + \mu_R}{\sqrt{\sigma_Q^2 + \sigma_R^2}}\right)$$

2.6.3 The Log-Normal Distribution

Sometimes it is not the random variable X that follows a normal distribution, but rather the log-transformed value of X, i.e. $ln(X) \sim N(\alpha, \beta^2)$, where parameters α and β^2 represent the mean and variance of the log-transformed values of X. In such cases the random variable X is said to be log-normally distributed as $X \sim LN(\alpha, \beta^2)$. The log-normal distribution is often evoked when considering the product of several random variables.

The pdf of the log-normal distribution is defined as:

$$f(x) = \frac{1}{x\beta\sqrt{2\pi}} \exp\left[-\frac{1}{2}\left(\frac{ln(x) - \alpha}{\beta}\right)^2\right], 0 < x < \infty \quad (2.45)$$

Note that unlike the pdf for the normal distribution, the log-normal distribution is only defined for positive values of x. This frequently makes it a preferred choice in the technical literature where variables often cannot physically take on negative value.

The mean and variance of a log-normally distributed random variable X are given as follows:

$$E(X) = \exp\left[\alpha + \frac{1}{2}\beta^2\right] \quad (2.46)$$

$$V(X) = \left(\exp\left[\beta^2\right] - 1\right)\exp\left[2\alpha + \beta^2\right] \quad (2.47)$$

If the mean and variance of X are expressed in real-space as $E(X) = \mu$ and $V(X) = \sigma^2$, then the equivalent parameters α and β^2 for the log-transformed values of X can be described as:

$$\alpha = ln(\mu) - \frac{1}{2} ln\left[\frac{\sigma^2}{\mu^2} + 1\right]$$
$$\beta^2 = ln\left[\frac{\sigma^2}{\mu^2} + 1\right] \quad (2.48)$$

EXAMPLE 2.10 DETERMINE QUANTILE OF A LOG-NORMAL DISTRIBUTION

Assume X is a log-normal distributed random variable defined as $X \sim LN(\alpha, \beta^2)$. This means that the log-transformed version of X follows a normal distribution defined as:

$$ln\, X \sim N(\alpha, \beta^2)$$

The pth quantile x_p can now be found by solving Eq. 2.27 as:

$$P(X \le x_p) = p$$

To find the quantile, first convert from a log-normal to a normal distribution using the natural logarithm within the probability function as:

$$P(\ln X \le \ln x_p) = p$$

Next, the expression is reduced to a standard normal distribution using Eq. 2.40 as:

$$\Phi\left(\frac{\ln x_p - \alpha}{\beta}\right) = p$$

The log-transformed quantile is found by inverting the above expression, i.e. solving Eq. 2.28:

$$\ln x_p = \alpha + \beta\Phi^{-1}(p)$$

Finally, the quantile is derived by using an exponential function as:

$$x_p = \exp\left(\alpha + \beta\Phi^{-1}(p)\right)$$

The inverse (or quantile) of the standard normal distribution $\Phi^{-1}(p)$ can be evaluated using the EXCEL function *NORM.INV* as discussed in Example 2.8.

EXAMPLE 2.11 FACTOR OF SAFETY

A factor of safety F is defined as the ratio between resistance R and the load Q as $F = R / Q$. As long as $F > 1$ the structure is safe, whereas values of $F < 1$ imply a failure (load exceeds resistance). Assuming that both R and Q are log-normal distributed random variables, then $\ln R$ and $\ln Q$ are both normally distributed. Log-transforming the factor of safety gives:

$$\ln F = \ln R - \ln Q$$

As both $\ln R$ and $\ln Q$ are normally distributed and independent, it follows from Eq. 2.43 that $\ln F$ is also normally distributed with parameters:

$$\ln F \sim N\left(\alpha_R - \alpha_Q, \beta_R^2 + \beta_Q^2\right)$$

The probability of failure can now be calculated as:

$$\begin{aligned} P(\text{failure}) &= P(F<1) = P(\ln F<0) \\ &= \Phi\left(\frac{0-\alpha_R+\alpha_Q}{\sqrt{\beta_R^2+\beta_Q^2}}\right) \\ &= \Phi\left(\frac{-\ln\left[\frac{\mu_R}{\mu_Q}\sqrt{\left(1+CV_Q^2\right)/\left(1+CV_R^2\right)}\right]}{\sqrt{\ln\left[\left(1+CV_Q^2\right)/\left(1+CV_R^2\right)\right]}}\right) \end{aligned}$$

where the last expression is obtained by substituting the values of α and β^2 from Eq. 2.48, and CV is the coefficient of variation defined in section 2.5.3 as $CV = \sigma/\mu$.

2.6.4 Gamma and Exponential Distributions

The gamma distribution is often used to describe aspects of waiting time between events or the lifetime of components. The gamma distribution has two parameters, a shape parameter α and a scale parameter λ, and a pdf defined as:

$$f(x) = \frac{\lambda^\alpha}{\Gamma(\alpha)} x^{\alpha-1} \exp(-\lambda x), \text{ for } x \geq 0 \text{ and } 0 \text{ elsewhere} \quad (2.49)$$

where Γ is the gamma function defined as:

$$\Gamma(x) = \int_0^\infty u^{x-1} \exp(-u) du, \; x \geq 0 \quad (2.50)$$

The mean and variance of the gamma distribution are defined as

$$E(X) = \frac{\alpha}{\lambda} \quad (2.51)$$

$$V(X) = \frac{\alpha}{\lambda^2} \tag{2.52}$$

The special case where the shape parameter $\alpha = 1$ is an exponential distribution with pdf

$$f(x) = \lambda \exp(-\lambda x), x \geq 0 \text{ and } 0 \text{ otherwise} \tag{2.53}$$

The exponential distribution was discussed in Examples 2.3, 2.5, and 2.6.

2.6.5 Extreme Value Distributions

In civil engineering it is often the most extreme events that are of interest for design, e.g. designing structures to protect against extreme flood events or designing a heating system to ensure comfort during extreme cold spells. In cases where a dataset consists of the maximum (or minimum) values, it is common to use extreme value distributions. An example of an extreme value distribution is the Gumbel distribution.

The pdf and cdf of the Gumbel distribution are defined as:

$$f(x) = \frac{1}{\alpha} \exp\left[-\frac{x-\mu}{\alpha} - \exp\left(-\frac{x-\mu}{\alpha}\right)\right] \tag{2.54}$$

$$F(x) = \exp\left[-\exp\left(-\frac{x-\mu}{\alpha}\right)\right] \tag{2.55}$$

The mean and variance of the Gumbel distribution are defined as:

$$E(X) = \mu + 0.5772\alpha \tag{2.56}$$

$$V(X) = \alpha^2 \frac{\pi^2}{6} \tag{2.57}$$

Extreme value distributions are often used to calculate design events, defined as quantiles in the extreme value distribution, x_T, where the subscript T is a specific return period and measured in time units (commonly years) but, critically, defined directly from a probability as:

$$T = \frac{1}{1 - F(x_T)} \tag{2.58}$$

Values of T are selected to reflect the acceptable level of probability of failure. For example, low impact failures such as sewer overflowing would typically consider return periods of around $T = 30$ years, whereas dam failure with potential loss of life would require design to much higher return periods such as $T = 10{,}000$ years.

EXAMPLE 2.12 CALCULATE THE 50-YEAR WIND SPEED

Based on historical observations it has been established that a Gumbel distribution representing the annual maximum daily mean wind speed (m/s) at a location is given as:

$$F(x) = \exp\left(-\exp\left(-\frac{x-6.25}{0.85}\right)\right)$$

Calculating the daily mean wind speed corresponding to a return period of $T = 50$ years requires solving Eq. 2.58 for x_{50}, which is essentially the same as finding the $(1-1/50)$ quantile of the Gumbel distribution, i.e. reorganising Eq. 2.58 as:

$$F(x_{50}) = 1 - \frac{1}{50}$$

The 50-year wind speed is therefore given as:

$$x_{50} = F^{-1}\left(1-\frac{1}{50}\right)$$

which for the Gumbel distribution gives:

$$x_{50} = 6.25 - 0.85\, ln\left(-ln\left[1-\frac{1}{50}\right]\right) = 9.57\,\text{m/s}$$

2.6.6 The Bernoulli Distribution

The Bernoulli distribution can be used to describe a discrete random variable that can take on two values: $X = 1$ with probability p or $X = 0$ with probability $1-p$. The probability mass function (pmf) for a Bernoulli distributed random variable is defined as:

$$f(x) = p^x (1-p)^{1-x},\ x = 0 \text{ or } 1 \tag{2.59}$$

The mean and variance of the Bernoulli distribution are defined as:

$$E(X) = p \tag{2.60}$$

$$V(X) = p(1-p) \tag{2.61}$$

2.6.7 The Binomial Distribution

A binomial distribution describes the case where n independent Bernoulli experiments are undertaken. The outcome of each experiment is either a "success" with probability p or a "failure" with probability $(1-p)$. A random variable X that counts the number of successes in the n experiments follows a binomial distribution $X \sim B(n,p)$. Obtaining $X = k$ successes in n experiments can be achieved in a number of ways. For example, obtaining exactly $k = 1$ successes in $n = 4$ experiments, where "success" is defined as 1 and "failure" defined as 0, can be achieved in four different ways as shown in Figure 2.5.

Each sequence consisting of exactly k successes <u>AND</u> $n-k$ failures has a probability of $p^k(1-p)^{n-k}$ according to Eq. 2.6. The total number of possible sequences representing k successes out of n experiments can be calculated using the binomial coefficient:

$$\binom{n}{k} = \frac{n!}{k!(n-k)!} \tag{2.62}$$

Each of the $\binom{n}{k}$ sequences has an equal probability of occurring, and since the probability of k successes is obtained in either sequence 1 <u>OR</u> sequence 2 <u>OR</u> ... <u>OR</u> sequence $\binom{n}{k}$, the total probability of $X = k$ (and therefore the probability mass function) is defined by Eq. 2.3 as the sum of $\binom{n}{k}$ independent terms as:

$$f(k) = \underbrace{p^k(1-p)^{n-k} + \ldots + p^k(1-p)^{n-k}}_{\binom{n}{k} \text{ independent terms}} \tag{2.63}$$

which results in a pmf defined as:

$$f(k) = \binom{n}{k} p^k (1-p)^{n-k}, \; k = 0,1,\ldots,n \tag{2.64}$$

The mean and variance of the binomial distribution are defined as:

$$E(X) = np \tag{2.65}$$

$$V(X) = np(1-p) \tag{2.66}$$

Experiment 1:	1	0	0	0
Experiment 2:	0	1	0	0
Experiment 3:	0	0	1	0
Experiment 4:	0	0	0	1

Figure 2.5 Four ways of obtaining $k = 1$ "successes" in $n = 4$ experiments.

EXAMPLE 2.13 QUALITY CONTROL OF CONCRETE ELEMENTS

Five concrete elements are needed to design a floor structure. The manufacturer delivers the elements and specifies that the probability of a faulty element is 0.02. What is the probability of not being able to build the floor if $n = 5$ elements are selected at random?

Answer: Consider the selection of each individual element as a Bernoulli experiment with two outcomes: "success" when the element is not faulty (probability 0.98) and "failure" if the element is faulty (probability 0.02). Let X be a random variable representing the number of successes (not-faulty elements); then the probability of k successes is defined according to the binomial pmf as

$$P(X=k)=f(k)=\binom{5}{k}0.98^{k}\times 0.02^{5-k}, k=0,1,2,3,4,5$$

The probability calculated for each k is shown in the table below as well as the outcome in terms of being able to construct or not.

Satisfactory, k	*Failure*, $n-k$	$f(x)$	*Outcome*
0	5	0.0000	No construction
1	4	0.0000	No construction
2	3	0.0001	No construction
3	2	0.0038	No construction
4	1	0.0922	No construction
5	0	0.9039	Construction

In this case the floor construction can only go ahead if all of the elements are not faulty, i.e. if $k = 5$ successes. Therefore, the probability of not being able to construct the floor is:

$$P(\text{failure})=P(X<5)=1-P(X=5)=1-f(X=5)$$
$$=1-0.9039=0.0961$$

If this probability is considered unacceptably high, then there is always the option to acquire more elements. For example, buying an additional element increases the pool to $n = 6$ elements. The probabilities for different values of k for $n = 6$ are described using a binomial distribution:

$$f(k)=\binom{6}{k}0.98^{k}\times 0.02^{6-k}, k=0,1,2,3,4,5,6$$

and the results shown in the table below.

Satisfactory, k	*Failure, n − k*	$f(x)$	*Outcome*
0	6	0.0000	No construction
1	5	0.0000	No construction
2	4	0.0000	No construction
3	3	0.0002	No construction
4	2	0.0055	No construction
5	1	0.1085	Construction
6	0	0.8858	Construction

In this case, construction can go ahead if even if one faulty element is included in the six. Therefore, the probability of not being able to construct the floor is

$$P(\text{failure}) = P(X < 5) = 1 - \left(P(X = 5) + P(X = 6)\right) = 1 - \left(f(5) + f(6)\right)$$
$$= 1 - (0.1085 + 0.8858) = 0.0057 < 0.0961$$

Of course, this reduction in probability comes at the expense of having to acquire an additional element.

Calculation of $P(X < 5) = P(X \leq 4)$ can also be calculated directly using EXCEL's *BINOM.DIST* function as shown in the figure below.

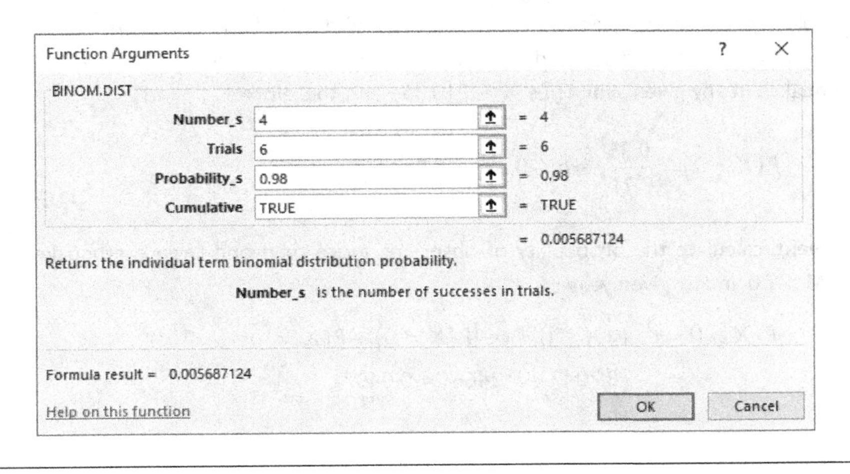

2.6.8 The Poisson Distribution

The Poisson distribution describes the probability of observing the number of occurrences of an event within a time or space interval of length t. Examples include the number of earthquakes in a year or the number of cars crossing a bridge in an hour. A Poisson distributed random variable X that describes the probability of observing k instances of an event in a time interval t has a pmf defined as:

$$P(X=k)=\frac{\lambda^k}{k!}exp(-\lambda),\ k=0,1,2, \quad (2.67)$$

where λ is the model parameter and represents the average occurrence rate of events in the time interval (e.g. average number of earthquakes per year). The short notation is $X \sim Pois(\lambda)$. The mean and variance of the Poisson distribution are given as

$$E(X)=\lambda \quad (2.68)$$

$$V(X)=\lambda \quad (2.69)$$

Values of the cumulative distribution function $P(X \leq k)$ can be calculated using the EXCEL function *POISSON.DIST*.

EXAMPLE 2.14 RISK OF EARTHQUAKES

Large earthquakes defined as $M > 7.0$ on the Richter scale occur at an average rate of $\lambda = 0.35$ year^{-1} in an area of the western Pacific. Calculate the probability of observing exactly two earthquakes with $M > 7.0$ in any given year.

Answer: First define a random variable X representing the number of earthquakes in any given year. Thus $X \sim P(0.35)$, and therefore:

$$P(X=2)=\frac{0.35^2}{2!}exp(-0.35)=0.0432$$

Next, calculate the probability of observing more than one large earthquake, $M > 7.0$, in any given year:

$$P(X>1)=1-P(X\leq 1)=1-[P(X=0)+P(X=1)]$$
$$=1-(0.7047+0.2466)=0.0487$$

Consider two Poisson distributed and independent random variables X and Y such that $X \sim Pois(\lambda)$ and $Y \sim Pois(\nu)$. The sum of the random variables $X+Y$ is itself following a Poisson distribution defined as:

$$X+Y \sim Pois(\lambda+\nu) \quad (2.70)$$

EXAMPLE 2.15 ARRIVAL OF CARS AND LORRIES

A traffic survey is counting the number of cars and lorries crossing a bridge per minute during rush hour. The arrival of cars and lorries is assumed independent and described by two Poisson distributed random variables:

X : number of cars arriving within a minute

Y : number of lorries arriving within a minute

The average rate of arrival for cars is $\lambda = 3$ and for lorries $\nu = 2$ such that $X \sim Pois(\lambda)$ and $Y \sim Pois(\nu)$. Find the probability that the total number of vehicles (cars + lorries) crossing the bridge exceeds ten, at which point congestion will start.

First, the total number of vehicles arriving follows a Poisson distribution as: $X + Y \sim Pois(3 + 2)$. The probability of $X + Y > 10$ is therefore:

$$P(X + Y > 10) = 1 - P(X + Y \leq 10)$$

The probability $P(X + Y \leq 10)$ can be evaluated by summing all the individual contributions $P(X + Y = 0) + P(X + Y = 1) + \ldots + P(X + Y = 10)$. Alternatively, the terms can be found using the *POISSON.DIST* function in EXCEL (see figure below). Thus:

$$P(X + Y > 10) = 1 - P(X + Y \leq 10) = 1 - 0.9863 = 0.0137$$

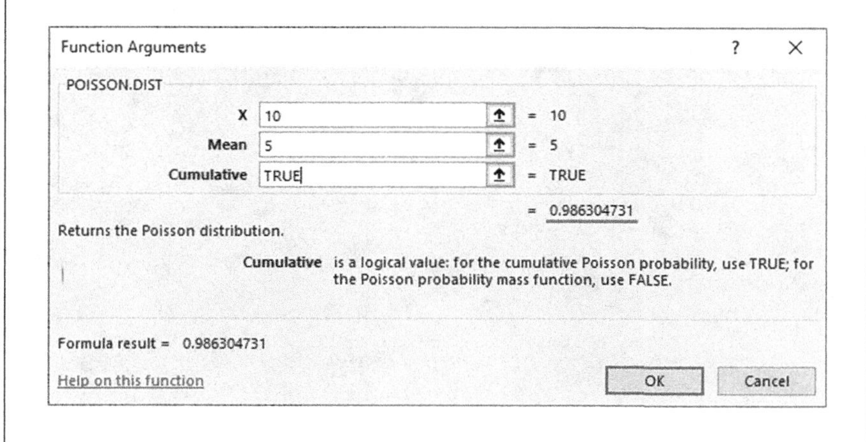

Appendix A: Cumulative Standard Normal Distribution

Tabulated values of $P(Z \leq z)$ where $Z \sim N(0,1)$

z	*0*	*0.01*	*0.02*	*0.03*	*0.04*	*0.05*	*0.06*	*0.07*	*0.08*	*0.09*
0	0.5	0.50399	0.50798	0.51197	0.51595	0.51994	0.52392	0.5279	0.53188	0.53586
0.1	0.53983	0.5438	0.54776	0.55172	0.55567	0.55962	0.56356	0.56749	0.57142	0.57535
0.2	0.57926	0.58317	0.58706	0.59095	0.59483	0.59871	0.60257	0.60642	0.61026	0.61409
0.3	0.61791	0.62172	0.62552	0.6293	0.63307	0.63683	0.64058	0.64431	0.64803	0.65173
0.4	0.65542	0.6591	0.66276	0.6664	0.67003	0.67364	0.67724	0.68082	0.68439	0.68793
0.5	0.69146	0.69497	0.69847	0.70194	0.7054	0.70884	0.71226	0.71566	0.71904	0.7224
0.6	0.72575	0.72907	0.73237	0.73565	0.73891	0.74215	0.74537	0.74857	0.75175	0.7549
0.7	0.75804	0.76115	0.76424	0.7673	0.77035	0.77337	0.77637	0.77935	0.7823	0.78524
0.8	0.78814	0.79103	0.79389	0.79673	0.79955	0.80234	0.80511	0.80785	0.81057	0.81327
0.9	0.81594	0.81859	0.82121	0.82381	0.82639	0.82894	0.83147	0.83398	0.83646	0.83891
1	0.84134	0.84375	0.84614	0.84849	0.85083	0.85314	0.85543	0.85769	0.85993	0.86214
1.1	0.86433	0.8665	0.86864	0.87076	0.87286	0.87493	0.87698	0.879	0.881	0.88298
1.2	0.88493	0.88686	0.88877	0.89065	0.89251	0.89435	0.89617	0.89796	0.89973	0.90147
1.3	0.9032	0.9049	0.90658	0.90824	0.90988	0.91149	0.91309	0.91466	0.91621	0.91774
1.4	0.91924	0.92073	0.9222	0.92364	0.92507	0.92647	0.92785	0.92922	0.93056	0.93189
1.5	0.93319	0.93448	0.93574	0.93699	0.93822	0.93943	0.94062	0.94179	0.94295	0.94408
1.6	0.9452	0.9463	0.94738	0.94845	0.9495	0.95053	0.95154	0.95254	0.95352	0.95449
1.7	0.95543	0.95637	0.95728	0.95818	0.95907	0.95994	0.9608	0.96164	0.96246	0.96327
1.8	0.96407	0.96485	0.96562	0.96638	0.96712	0.96784	0.96856	0.96926	0.96995	0.97062
1.9	0.97128	0.97193	0.97257	0.9732	0.97381	0.97441	0.975	0.97558	0.97615	0.9767
2	0.97725	0.97778	0.97831	0.97882	0.97932	0.97982	0.9803	0.98077	0.98124	0.98169
2.1	0.98214	0.98257	0.983	0.98341	0.98382	0.98422	0.98461	0.985	0.98537	0.98574
2.2	0.9861	0.98645	0.98679	0.98713	0.98745	0.98778	0.98809	0.9884	0.9887	0.98899

(Continued)

z	0	0.01	0.02	0.03	0.04	0.05	0.06	0.07	0.08	0.09
2.3	0.98928	0.98956	0.98983	0.9901	0.99036	0.99061	0.99086	0.99111	0.99134	0.99158
2.4	0.9918	0.99202	0.99224	0.99245	0.99266	0.99286	0.99305	0.99324	0.99343	0.99361
2.5	0.99379	0.99396	0.99413	0.9943	0.99446	0.99461	0.99477	0.99492	0.99506	0.9952
2.6	0.99534	0.99547	0.9956	0.99573	0.99585	0.99598	0.99609	0.99621	0.99632	0.99643
2.7	0.99653	0.99664	0.99674	0.99683	0.99693	0.99702	0.99711	0.9972	0.99728	0.99736
2.8	0.99744	0.99752	0.9976	0.99767	0.99774	0.99781	0.99788	0.99795	0.99801	0.99807
2.9	0.99813	0.99819	0.99825	0.99831	0.99836	0.99841	0.99846	0.99851	0.99856	0.99861
3	0.99865	0.99869	0.99874	0.99878	0.99882	0.99886	0.99889	0.99893	0.99896	0.999
3.1	0.99903	0.99906	0.9991	0.99913	0.99916	0.99918	0.99921	0.99924	0.99926	0.99929
3.2	0.99931	0.99934	0.99936	0.99938	0.9994	0.99942	0.99944	0.99946	0.99948	0.9995
3.3	0.99952	0.99953	0.99955	0.99957	0.99958	0.9996	0.99961	0.99962	0.99964	0.99965
3.4	0.99966	0.99968	0.99969	0.9997	0.99971	0.99972	0.99973	0.99974	0.99975	0.99976
3.5	0.99977	0.99978	0.99978	0.99979	0.9998	0.99981	0.99981	0.99982	0.99983	0.99983
3.6	0.99984	0.99985	0.99985	0.99986	0.99986	0.99987	0.99987	0.99988	0.99988	0.99989
3.7	0.99989	0.9999	0.9999	0.9999	0.99991	0.99991	0.99992	0.99992	0.99992	0.99992
3.8	0.99993	0.99993	0.99993	0.99994	0.99994	0.99994	0.99994	0.99995	0.99995	0.99995
3.9	0.99995	0.99995	0.99996	0.99996	0.99996	0.99996	0.99996	0.99996	0.99997	0.99997
4	0.99997	0.99997	0.99997	0.99997	0.99997	0.99997	0.99998	0.99998	0.99998	0.99998

Appendix B: Summary of Distributions

Summary of distributions introduced in the text.

Distribution	*Pdf*	*Mean*	*Variance*
Continuous distributions			
Uniform	$\frac{1}{b-a}$	$\frac{a+b}{2}$	$\frac{(b-a)^2}{12}$
Normal	$\frac{1}{\sqrt{2\pi}}\frac{1}{\sigma}\exp\left[-\frac{1}{2}\left(\frac{x-\mu}{\sigma}\right)^2\right]$	μ	σ^2
Log-normal	$\frac{1}{x\beta\sqrt{2\pi}}\exp\left[-\frac{1}{2}\left(\frac{Ln(x)-\alpha}{\beta}\right)^2\right]$	$\exp\left[\alpha+\frac{1}{2}\beta^2\right]$	$\left(\exp\left[\beta^2\right]-1\right)\exp\left[2\alpha+\beta^2\right]$
Exponential	$\lambda\exp(-\lambda x)$	$\frac{1}{\lambda}$	$\frac{1}{\lambda^2}$
Gamma	$\frac{\lambda^\alpha}{\Gamma(\alpha)}x^{\alpha-1}\exp(-\lambda x)$	$\frac{\alpha}{\lambda}$	$\frac{\alpha}{\lambda^2}$
Gumbel	$\frac{1}{\alpha}\exp\left[-\frac{x-\mu}{\alpha}-\exp\left(-\frac{x-\mu}{\alpha}\right)\right]$	$\mu+0.5772\alpha$	$\alpha^2\frac{\pi^2}{6}$
Discrete distributions			
Bernoulli	$p^x(1-p)^{1-x}$	p	$p(1-p)$
Binomial	$\binom{n}{k}p^k(1-p)^{n-k}$	np	$np(1-p)$
Poisson	$\frac{\lambda^k}{k!}\exp(-\lambda)$	λ	λ

Chapter 3

Parameter Estimation

3.1 FITTING STATISTICAL MODELS TO DATA

Each of the frequency distributions listed in Chapter 2 is characterised by one or more parameters. For example, the exponential distribution has a single parameter λ while the normal distribution has two parameters μ and σ^2. A typical problem is that for a particular phenomenon (e.g. maximum wind speed or material strength) a theoretical frequency distribution is known or assumed, but the parameters are unknown, and a best guess (or estimate) must be derived from a sample of observed data points $x_1, x_2, \ldots, x_n$. The term *sample* is used here to describe a set of data points or observations that represent a subset of all possible observations. For example, measuring the strength of ten concrete elements in a laboratory gives a sample size of $n = 10$. Testing an additional ten elements will increase the sample size to $n = 20$. In theory we could test an infinite number of elements ($n \to \infty$) to ensure we have complete knowledge of the distribution of strength and its parameters. However, in practice, only a limited number of experiments can be conducted, and therefore estimated values of the distribution parameters represent best guesses based on the sample available.

Several methods are available for fitting theoretical probability distributions to real-world data samples. The complexity of these methods varies and depends on the number and nature of assumptions that are made about the data and phenomena being modelled. In general, consider a distribution with a probability density function (pdf) $f(x)$ characterised by a set of k *unknown* model parameters $\theta_1, \ldots, \theta_k$. For example, the exponential distribution has $k = 1$ parameters λ.

In the following examples, two different methods will be introduced: the method of moments and maximum likelihood. Common for both methods is that a set of mathematical algorithms (*an estimator*) will be produced that allows a best guess of the parameter values to be calculated based on the observed data (*an estimate*).

DOI: 10.1201/9781032700373-3

3.2 THE METHOD OF MOMENTS

The concept of statistical moments (mean and variance) was introduced in section 2.5. It was shown how a theoretical frequency distribution is characterised by its statistical moments, and that these in turn depends on the as yet unknown model parameters. Next, the equivalent sample moments are derived from the observed data $x_1, x_2, \ldots, x_n$. Here the term *sample* refers to numbers derived from the finite observed data as discussed above. The sample mean $\overline{x}$ and sample variance s^2 are defined as:

$$\begin{aligned} \overline{x} &= \frac{1}{n}\sum_{i=1}^{n} x_i \\ s^2 &= \frac{1}{n-1}\sum_{i=1}^{n}(x_i - \overline{x})^2 \end{aligned} \tag{3.1}$$

Finally, to find the values of the distribution parameters, equate the theoretical statistical moments with the equivalent sample moments and find the parameter values that solves this set of equations. If a distribution has more than two parameters, then higher-order sample and theoretical moments need to be considered, which can become very complex.

EXAMPLE 3.1 ESTIMATING PARAMETERS OF THE GUMBEL DISTRIBUTION: PART 1

Consider a Gumbel distribution with a pdf defined as:

$$f(x) = \frac{1}{\alpha}\exp\left[-\frac{x-\mu}{\alpha} - \exp\left[-\frac{x-\mu}{\alpha}\right]\right]$$

According to section 2.5 the theoretical moments of the Gumbel distribution are defined as:

$$E(X) = \int_{-\infty}^{\infty} xf(x)dx = \alpha + 0.5772$$

$$V(X) = \int_{-\infty}^{\infty} (x-\mu)^2 f(x)dx = \alpha^2\frac{\pi^2}{6}$$

Equating $E(X)$ and $V(X)$ with the corresponding two sample moments from Eq. 3.1 gives:

$$\overline{x} = \mu + 0.5772\alpha$$

$$s^2 = \alpha^2\frac{\pi^2}{6}$$

which upon rearrangement gives the estimators:

$$\hat{\alpha} = \frac{\sqrt{6}}{\pi} s$$

$$\hat{\mu} = \overline{x} - 0.5772\alpha$$

The "hat" notation " $\hat{}$ " is used to indicate an estimator or estimated value, as opposed to a theoretical value.

EXAMPLE 3.2 ESTIMATING PARAMETERS OF THE GUMBEL DISTRIBUTION: PART 2

The River Avon drains a catchment area of approximately 1500 km^2 in the west of England before it enters the city of Bath. A gauging station installed on the river has been recording river flow (m^3/s) continuously every 15 minutes since 1969. A study to understand the risk and uncertainty of flooding to the city, the single largest 15-minute flow observation is extracted for each hydrological year, defined as the period starting 1 October and running until 30 September the following year. The hydrological year is preferred over the calendar year (1 January to 31 December) to better align with the natural hydrological cycle of northwest Europe (wet winters and dry summers).

Extracting the annual maximum 15-minute flow observation for each hydrological year results in an annual maximum series of peak flow consisting of 51 observations covering the period 1969–2020. These data can be accessed via the National River Flow Archive (NRFA), and a time series plot of the data is shown in the figure below, highlighting the random year-to-year variability of the data.

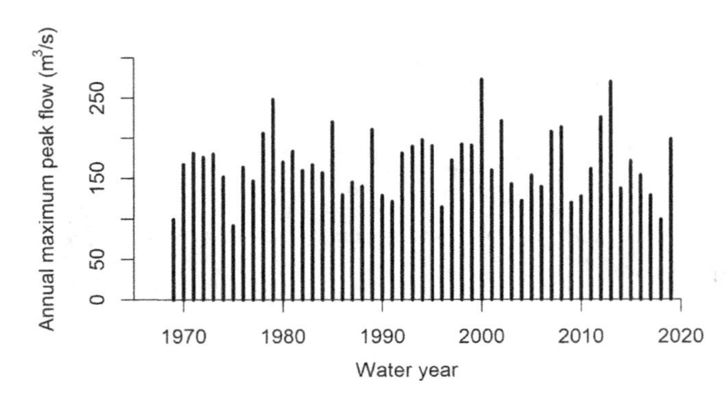

Time series plot of annual maximum series of peak flow from the River Avon.

Using Eq. 3.1 the sample mean and standard deviation of the 51 observations (the sample) are calculated:

$$\bar{x} = 168.6\,\mathrm{m^3/s}$$

$$s = 41.1\,m^3/s$$

Inserting these sample moments into the two parameter estimators derived in Example 3.1 gives:

$$\hat{\alpha} = \frac{\sqrt{6}}{\pi}s = \frac{\sqrt{6}}{\pi}41.1 = 32.05$$

$$\hat{\mu} = \bar{x} - 0.5772\alpha = 168.6 - 0.5772 \times 32.05 = 144.91$$

3.3 THE MAXIMUM LIKELIHOOD METHOD

The maximum likelihood method is an alternative method for estimating parameters of probability distributions. The maximum likelihood method is sometimes preferred as it can produce more precise estimates, and can be used more flexibly in non-standard situations, but this is at the expense of more complex calculations, often involving numerical optimisation.

Assume that the observations $x_1,\ldots,x_n$ are independent and follow a probability distribution with a pdf denoted $f(x)$. The pdf has a number, k, of as yet unknown parameters denoted by a vector θ. Typically, the number of parameters is equal to $k = 1$ or $k = 2$, but higher numbers are sometimes encountered. For example, the one-parameter exponential distribution has one parameter $\theta = \lambda$, while the log-normal distribution has two parameters $\theta = (\alpha, \beta^2)$. Next, consider the probability (or likelihood L) that *this* particular set of observations (sample) came to pass is the probability of $X = x_1$ AND probability of $X = x_2$ AND … AND probability of $X = x_n$ which is defined as:

$$L = P(X = x_1 \cap X = x_2 \cap \cdots \cap X = x_n) \tag{3.2}$$

Using that $P(X = x_i)$ is proportional to the pdf $f(x_i)$ (Eq. 2.8) and using the law of multiplying probabilities of independent events (Eq. 2.6) gives the likelihood function:

$$L(\theta) = \prod_{i=1}^{n} f(x_i) \tag{3.3}$$

The notation $L(\theta)$ is to emphasise that the likelihood function depends on one or more unknown parameters θ that can now be determined as the set of parameters that maximises the likelihood function L. Often it is more convenient to optimise the natural logarithm of the likelihood function, i.e.:

$$ln\, L(\theta) = ln\left[\prod_{i=1}^{n} f(x_i)\right] = \sum_{i=1}^{n} ln\, f(x_i) \tag{3.4}$$

The maximum likelihood estimate of the parameter is therefore the value that solves the equation:

$$\frac{\partial\, ln\, L(\theta)}{\partial \theta_i} = 0, i = 1, \ldots, k \tag{3.5}$$

The partial differentiation is used as some distributions have more than one parameter, in which case Eq. 3.5 must be solved for each parameter. For more complex distributions with two or more parameters, or when an analytical solution is not available, numerical optimisation methods might have to be used.

EXAMPLE 3.3 MAXIMUM LIKELIHOOD ESTIMATION OF PARAMETERS OF EXPONENTIAL DISTRIBUTION: PART I

The exponential distribution has a pdf defined as:

$$f(x) = \lambda exp(-\lambda x)$$

For a set of data $x_1, x_2, \ldots, x_n$ the likelihood function is defined as:

$$L = \prod_{i=1}^{n} f(x_i) = \prod_{i=1}^{n} \lambda exp(-\lambda x_i)$$

In this case it is more convenient to work with the log-likelihood function:

$$lnL = \sum_{i=1}^{n} lnf(x_i) = \sum_{i=1}^{n} ln[\lambda exp(-\lambda x_i)] = nln\lambda - \sum_{i=1}^{n} \lambda x_i$$

The optimal value of $\ln L$ is obtained when $d(\ln L) / d\lambda = 0$:

$$\frac{d(lnL)}{d\lambda} = \frac{n}{\lambda} - \sum_{i=1}^{n} x_i = 0$$

and finally:

$$\hat{\lambda} = \frac{n}{\sum_{i=1}^{n} x_i} = \frac{1}{\overline{x}}$$

Therefore, the optimal parameter value is the inverse of the mean of the data.

EXAMPLE 3.4 MAXIMUM LIKELIHOOD ESTIMATION OF PARAMETERS OF EXPONENTIAL DISTRIBUTION: PART 2

Estate management has recorded the lifetime of newly installed lightbulbs across their buildings. They recorded the lifetime (in years) of 20 lightbulbs from the time of installation installed until recorded as broken. The recorded lifetimes are shown in the table below.

Lightbulb no	*Lifetime (years)*	*Lightbulb no*	*Lifetime (years)*	*Lightbulb no*	*Lifetime (years)*	*Lightbulb no*	*Lifetime (years)*
1	0.14	6	2.87	11	1.80	16	0.49
2	0.73	7	0.10	12	2.47	17	7.44
3	0.77	8	0.36	13	1.43	18	0.36
4	0.86	9	0.57	14	2.85	19	1.84
5	0.71	10	4.83	15	3.20	20	1.79

Denoting the lifetime as a random variable X, the sample mean of the recorded lifetimes is calculated using Eq. 3.1 as:

$$\overline{x} = \frac{1}{20}(0.14 + 0.73 + \cdots + 1.84 + 1.79) = 1.78 \text{ years}$$

Assuming the lifetime of a lightbulb follows an exponential distribution, then the parameter λ can be estimated as:

$$\lambda = \frac{1}{\overline{x}} = \frac{1}{1.78} = 0.56$$

Thus, the pdf of lifetime of lightbulbs is:

$$f(x) = 0.56 exp(-0.56x), \text{ for } x > 0 \text{ and } 0 \text{ elsewhere}$$

In general, the maximum likelihood method is more accurate than the method of moments, but it also has some practical drawbacks, including increased complexity and the fact that sometimes no numerical solution can be found. But maximum likelihood is also a more flexible approach that can be tailored to non-standard datasets, such as the one in Example 3.5.

EXAMPLE 3.5 MAXIMUM LIKELIHOOD ESTIMATION FOR CENSORED DATA

Consider the n "experiments" that generated the n observations $x_1,\ldots,x_n$. For example, each observation could represent the strength of a concrete element tested under identical laboratory conditions, and a total of n elements were tested. The maximum strength that can be measured by the testing equipment is denoted x_0. If the maximum strength has been reached without breaking the element, the data are censored as we no longer know the exact strength but only know that it must be bigger than the upper measuring limit of x_0. Assume that in m out of the total of n experiments, the maximum capacity was exceeded. For this case, the likelihood function, L, needs to be redefined accordingly. For this case we have m observations where the strength exceeds x_0, and thus $n-m$ observations where the equipment was able to provide an accurate measurement. These two conditions need to be represented differently in the likelihood function as:

$$L = \underbrace{\prod_{i=1}^{n-m} P(X = x_i)}_{X \leq x_0} \times \underbrace{\prod_{j=1}^{m} P(X > x_0)}_{X \leq x_0}$$

This expression can be rewritten in terms of the pdf, $f(x)$, and cdf, $F(x)$, of the chosen distribution as:

$$L = \underbrace{\prod_{i=1}^{n-m} f(x_i)}_{X \leq x_0} \times \underbrace{\prod_{j=1}^{m} \left(1 - F(x_0)\right)}_{X > x_0}$$

Finally, the log-likelihood function is derived as:

$$\ln L = \sum_{i=1}^{n-m} \ln f(x_i) + \sum_{j=1}^{m} \ln\left[1 - F(x_0)\right]$$

This type of estimator can be used if, for example, the lightbulb data in Example 3.4 were capped at a maximum of 2 years, meaning if the lifetime exceeded 2 years, then a maximum value of 2.00 years was recorded.

Appendix: Summary of Parameter Estimators

A summary of method of moment parameter estimators for distributions discussed in Chapter 2. For some distributions the maximum likelihood (M-L) gives simpler estimators, for some this is the case for the method of moments (MoM). The table includes the simplest estimator with an indication of the underlying method.

Assume a sample of n data points are available $x_1, x_2, \ldots, x_n.$, then the following mathematical notations are used in the estimators.

$x_{[1]}$	Largest observation, $max(x_1, x_2, \ldots, x_n.)$
$x_{[n]}$	Smallest of the n observation, $min(x_1, x_2, \ldots, x_n.)$
$\overline{x}$	Sample mean
s	Sample standard deviation

Table Summary of parameter estimation equations

Distribution	*pdf*	*Unknown parameters*	*Method*	*Estimators*
Continuous distributions				
Uniform	$\frac{1}{b-a}$	(a,b)	M-L	$(x_{[1]}, x_{[n]})$
Exponential	$\lambda \exp(-\lambda x)$	λ	M-L, MoM	$\frac{1}{\bar{x}}$
Gamma	$\frac{1}{\Gamma(k)\lambda^k} x^{k-1} \exp(-\lambda x)$	(α,λ)	MoM	$\left(\frac{\bar{x}^2}{s^2}, \frac{\bar{x}}{s^2}\right)$
Normal	$\frac{1}{\sqrt{2\pi}\sigma} \exp\left[-\frac{1}{2}\left(\frac{x-\mu}{\sigma}\right)^2\right]$	(μ,σ^2)	M-L, MoM	$(\bar{x}, s^2)$
Log-normal	$\frac{1}{x\beta\sqrt{2\pi}} \exp\left[-\frac{1}{2}\left(\frac{\ln x-\alpha}{\beta}\right)^2\right]$	(α,β^2)	MoM	$\ln(\bar{x}) - \frac{1}{2}\ln\left[\frac{s^2}{\bar{x}^2}+1\right]$, $\ln\left[\frac{s^2}{\bar{x}^2}+1\right]$

(*Continued*)

Table Summary of parameter estimation equations

Distribution	*pdf*	*Unknown parameters*	*Method*	*Estimators*
Gumbel	$\frac{1}{\alpha}\exp\left[-\frac{x-\mu}{\alpha}-\exp\left[-\frac{x-\mu}{\alpha}\right]\right]$	(μ,α)	MoM	$\left(\bar{x}-0.5772\alpha,\frac{\sqrt{6}}{\pi}s\right)$
Discrete distributions				
Poisson	$\frac{\lambda^k}{k!}\exp(-\lambda)$	λ	ML, MoM	$\bar{x}$

Chapter 4

Data Visualisation

Before undertaking any statistical data analysis, it is always good practice to first examine the data through a visual assessment. This section presents three different types of plots that can be used for examining and presenting data: histograms, probability plots, and box plots. The chapter ends with a general discussion of good practice when creating data visualisations.

4.1 HISTOGRAMS

A histogram is a graphical summary of a dataset showing the central tendency, spread, and general shape of the dataset. Consider a dataset consisting of n observations. From this data a histogram is created by first subdividing the range spanned by the observations into a number of intervals. Next the number of observations that falls within each interval is counted. Finally, a rectangle is drawn over each interval where the height is defined as:

$$\frac{Number\ of\ observations\ in\ interval}{Total\ number\ of\ observations \times width\ of\ interval} \tag{4.1}$$

It is important to normalise the number of observations in each interval using the width of the interval to ensure that the total area of the histogram is 1, thereby allowing a direct comparison with the probability density function.

There are several rules of thumb for how to determine the number of intervals (or bins) denoted k. An example is Rice's formula:

$$k = 2n^{1/3} \tag{4.2}$$

where n is the total number of observations in the dataset.

Make sure the interval boundaries are defined by one more digit than the observations to ensure that there is no doubt in which interval an observation belongs.

DOI: 10.1201/9781032700373-4

EXAMPLE 4.1 HISTOGRAM OF ANNUAL RAINFALL IN OXFORD

Rainfall has been recorded at Heathrow Airport continuously since 1949, and annual rainfall totals (mm/year) are available for the period 70 years (1949–2017). An initial investigation of the dataset should include a time series plot as well as a histogram. Using Rice's formula gives $2 \times 70^{1/3} \approx 8$ bins. The figure below shows the time series plot (top), the number of data points in each interval (middle), and the actual histogram (bottom) where the y-axis indicates frequencies obtained using Eq. 4.1.

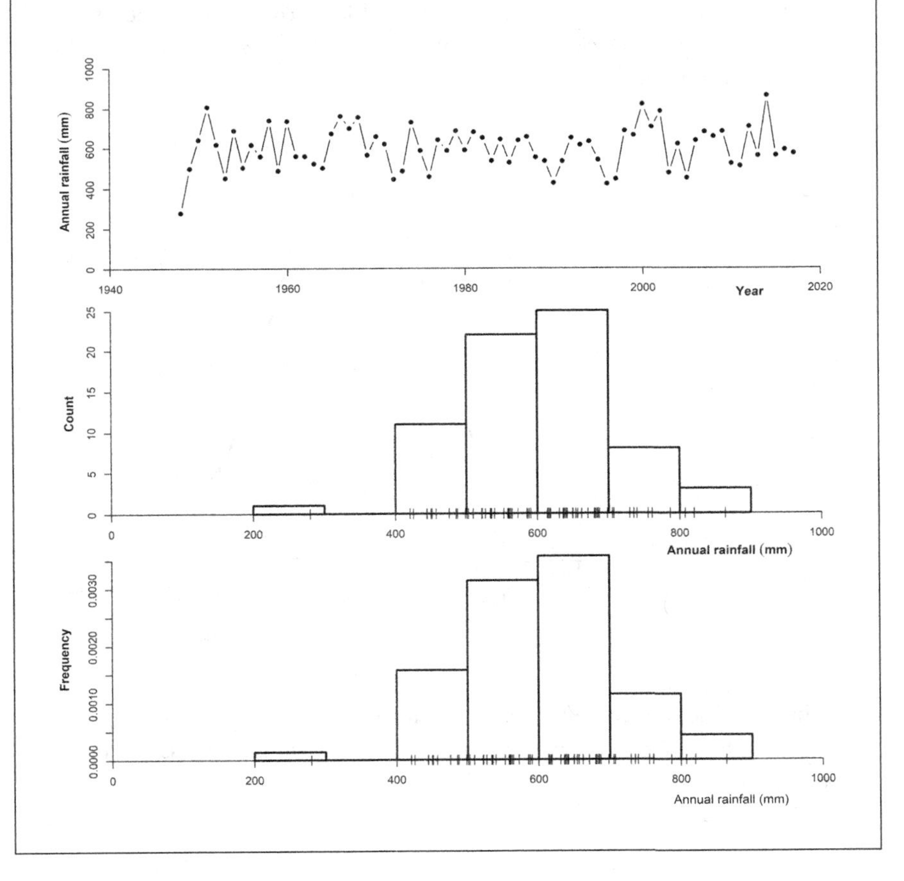

EXAMPLE 4.2 COMPARING HISTOGRAMS AND PDF

It is possible to compare a histogram constructed per Eq. 4.1 directly with a probability density function (pdf) for a given distribution. It is, however, not always easy to judge how well a particular distribution fits a dataset from such plot, nor to compare the relative merits of two different candidate distributions. Using the annual rainfall data from Example 4.1, the parameters of both the normal and the log-normal distributions are estimated using the method of moment as outlined in Chapter 3. Next, the two pdf functions are plotted on the same figure as the histogram of the annual rainfall. The solid line represents the pdf of the normal distribution, while the broken line represents the pdf of the log-normal distribution.

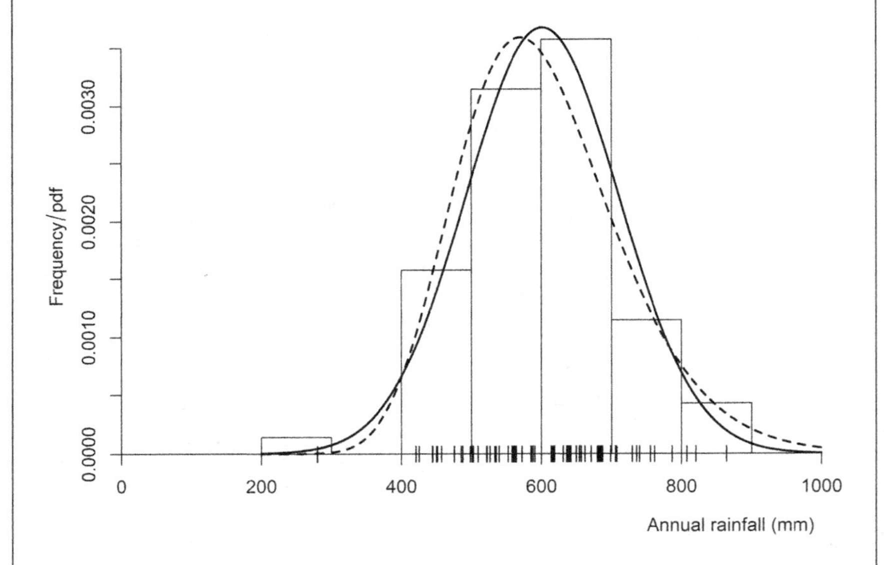

The figure shows that while both candidate distributions appear to give a reasonable fit to the histogram, it is not clear or obvious which is the more appropriate choice.

4.2 PROBABILITY PAPER

While histograms are useful tools for visualisation of data, it can be difficult to ascertain if a particular probability distribution fits the data well or not, as illustrated in Example 4.2. For this purpose, specially designed probability paper is more commonly used. Probability paper is a scatter plot where the co-ordinate system is designed to allow the cdf of a specific distribution to plot as straight line and allows for a direct comparison with the empirical non-exceedance probability of the data points.

Consider a cdf for a chosen distribution evaluated at a value x_p:

$$F(x_p) = p \tag{4.3}$$

where p is a probability and x_p is the pth quantile of the distribution which can be found by inverting the cdf as:

$$x_p = F^{-1}(p) \tag{4.4}$$

For some of the most commonly used distributions the function F^{-1} can be expressed as a linear function which allows easy-to-interpret plots to be created, as illustrated in the following examples.

EXAMPLE 4.3 CREATION OF GUMBEL PROBABILITY PAPER

The cdf of the Gumbel distribution is defined as:

$$F(x) = exp\left(-exp\left[-\frac{x-\mu}{\alpha}\right]\right)$$

The corresponding pth-quantile of the distribution is given as:

$$x_p = \mu - \alpha\, ln\left[-ln(-p)\right]$$

By defining $y_p = -ln\left[-ln(p)\right]$ the quantile x_p can be defined as a straight line in a (y_p, x_p) co-ordinate system as:

$$x_p = \mu + \alpha y_p$$

where μ and α are constant model parameters representing the intercept and slope of a straight line. Note that when working with the Gumbel distribution it is common to express the probability p as an equivalent report period $T = 1/(1-p)$.

EXAMPLE 4.4 CREATION OF NORMAL PROBABILITY PAPER

The quantile of a normally distributed random variable is, according to Eq. 2.42, defined as:

$$\Phi^{-1}(p) = \frac{x_p - \mu}{\sigma}$$

$$x_p = \mu + \sigma\Phi^{-1}(p)$$

Thus, the quantile x_p can be defined as a straight line in a $\left(\Phi^{-1}(p), x_p\right)$ coordinate system.

A summary of how to construct probability paper for common distributions is shown in Table 4.1.

The next step is to add the observations to the probability plot via the empirical non-exceedance probability. Each data point is assigned an estimated non-exceedance probability p using one of many available plotting positions. The term *empirical* is used here to emphasise that that the probability is derived without any assumption being made regarding the distribution of the data.

Consider a sample consisting of n observations $x_1,\ldots,x_n$. The sample is ranked in descending order, so that $x_{[1]} \geq \ldots \geq x_{[n]}$. The largest observation in the sample is assigned rank $r = 1$, the second largest $r = 2$, and so on, and the smallest $r = n$. Using only the rank and the total number of observations, the empirical non-exceedance probability $P\left(X \leq x_{[r]}\right)$ for each observation can be estimated as:

$$p_r = 1 - \frac{r}{n+1}, r = 1,\cdots,n \tag{4.5}$$

Table 4.1 Probability Paper for Common Distributions

Distribution	*cdf*	*Quantile*
Gumbel	$F\left(x_p\right) = exp\left(-exp\left[-\frac{x_p - \mu}{\alpha}\right]\right)$	$x_p = \mu - \alpha ln\left(-ln[p]\right)$
Normal	$\Phi\left(\frac{x-\mu}{\sigma}\right)$	$x_p = \mu + \sigma\Phi^{-1}(p)$
Log-normal	$\Phi\left(\frac{ln(x)-\mu}{\sigma}\right)$	$ln\left(x_p\right) = \mu + \sigma\Phi^{-1}(p)$

Next, the non-exceedance probability is transformed to conform to the specified probability paper. For example, for the Gumbel distribution the transformation is done as $y_p = -ln(-ln[p])$.

EXAMPLE 4.5 USING NORMAL PROBABILITY PAPER

This example illustrates how to create a probability plot for a normal distribution and use this plot to assess if the data follows a normal distribution.

The compressive strength of concrete made from a new formula is investigated by testing ten samples. The results are shown in the table below. While the same cement is used in each sample, the experiment gives different results each time the experiment is conducted. There might be many reasons for this such as slight differences in procedures used in each experiment, perhaps different technicians did the different test, or any other uncontrollable factors during the experiments (room temperature, degree of mixing of the concrete mix, etc.). These differences are all considered to be of a random nature.

Sample	*Strength (MPa)*	*Rank*	p_r	$\Phi^{-1}(p_r)$
1	33.2	9	0.1818	−0.9085
2	39.9	3	0.7273	0.6046
3	37.3	6	0.4545	−0.1142
4	38.1	4	0.6364	0.3488
5	40.5	2	0.8182	0.9085
6	31.8	10	0.0909	−1.3352
7	34.4	7	0.3636	−0.3488
8	33.8	8	0.2727	−0.6046
9	38.1	4	0.6364	0.3488
10	42.6	1	0.9091	1.3352

The following steps are required to create the probability plot. First, the rank of the observations is determined (column 3) and then the empirical non-exceedance probability p_r of each observation is defined using Eq. 4.5 (column 4).

Next, the non-exceedance probabilities are transformed in column 5 using the normal distribution as $\Phi^{-1}(p_r)$ (column 5). This transformation can be done using the EXCEL *NORM.INV* function.

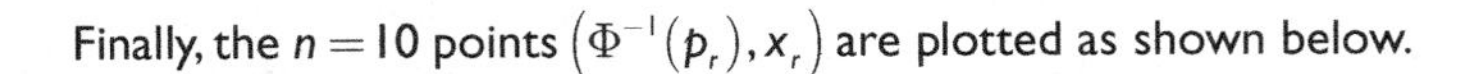
Finally, the $n = 10$ points $\left(\Phi^{-1}(p_r), x_r\right)$ are plotted as shown below.

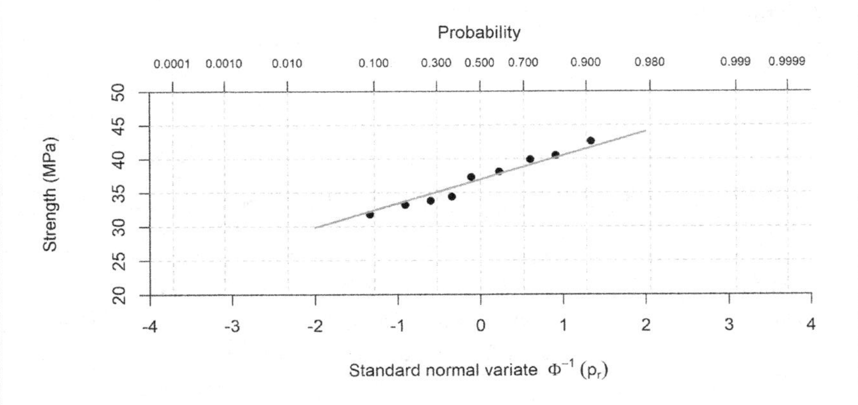

The grey line on the figure represents the cdf of the normal distribution and is defined using the mean and standard deviation of the material strength data as:

$$x_p = 36.97 + 3.54\Phi^{-1}(p)$$

The data points represent the ten observations and their empirical non-exceedance probability. As the data appears to follow a straight line defined by the normal distribution (the grey line), it is reasonable to assume that the data are normally distributed. More formal goodness-of-fit tests, such as the χ^2 test, can be used if further confirmation is required.

4.3 BOX PLOTS

Box plots, sometime known as box-whisker plots, are used to visualise the distribution of a dataset based on a five-number summary of the dataset consisting of the

- Minimum.
- 25% quantile.
- Median (50% quantile).
- 75% quantile.
- Maximum.

The minimum and maximum of a dataset can be extracted in EXCEL using the *MIN* and *MAX* functions. The quantiles can be extracted using the

QUANTILE function. The box plot is constructed as a rectangle spanning the 25% to the 75% quantile, with a horizontal line indicating the position of the median. A whisker is added to each end of the rectangle stretching from the 25% and 75% quantile to the minimum and maximum observation, respectively.

EXAMPLE 4.6 BOX PLOT OF ANNUAL RAINFALL TOTALS

A data series consisting of 70 years of annual rainfall totals (January–December) recorded at Heathrow Airport is shown in the top panel in the figure below. Next, a histogram of the data is shown in the middle panel. Using Rice's formula with $n = 70$ years of data results in eight bins. Finally, the box plot indicating the location of the 5-number summary of the data is shown.

Box plots contain less information than histograms, but they can be particularly useful when comparing the behaviour of two or more datasets. The figure

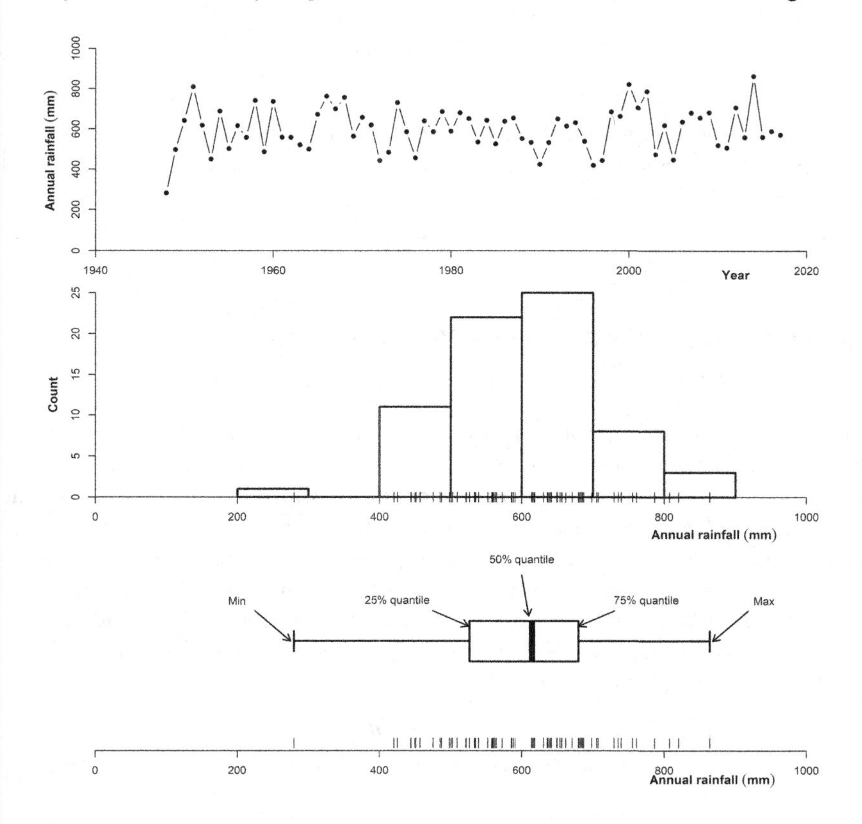

below shows a comparison of the distribution of annual rainfall totals (mm) recorded at Heathrow Airport in the South East of the UK, and at the village of Eskdalemuir in South West Scotland. In fact, even the driest year on record is more wet than the wettest year on record in London.

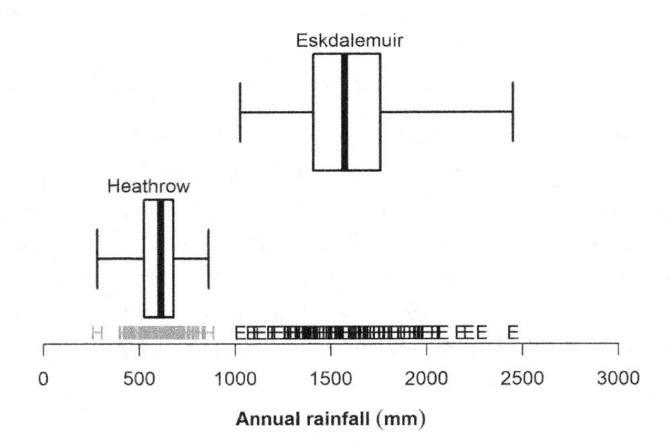

Looking at this figure it is very easy to compare the two datasets and see that rainfall in South West Scotland is higher than rainfall in the South East of the UK.

4.4 SOME GUIDANCE ON EFFECTIVE DATA VISUALISATION

The material is this section is mostly based on the seminal work published by Edward Tufte (Tufte, 2001) in his book *The Visual Display of Quantitative Information*. The overarching principle of statistical graphs according to Tufte is "Above all else show the data," i.e. graphs should communicate their message to the reader as simply as possible. To translate this mantra into more practical advice, Tufte advised some simple definitions and rules: primarily data ink is the amount of ink devoted to the non-redundant display of data, and the data-ink ratio is the ratio between the data and the total amount of ink used to create the graph.

$$Data-ink\ ration = \frac{Data\ ink}{Total\ ink\ used\ to\ create\ the\ figure} \tag{4.6}$$

The idea behind the data-ink ratio is to make designers of statistical graphs (i.e. you dear reader) consider if the figure communicates the data and findings as clearly as possible. Thus it is desirable to strive for a data-ink ratio as close to 1 as possible, whereas a ratio of zero would not show any of the data at all. Under the mantra of maximising the data-ink ratio consider erasing non-data ink and/or erasing redundant data-ink. Once finished, revise the figure and edit until satisfied that a good data-ink ratio has been achieved.

A good starting point is to write a mission-statement for your graph; what is this graph all about? What are you hoping to communicate? Below are some examples.

EXAMPLE 4.7 GRID LINES, GRID LINES EVERYWHERE

Plot values of the California Bearing Ratio (%) and plasticity index (%) obtained from laboratory experiments on 20 soil samples in the form of a simple x-y scatter diagram. The figures below show three attempts to create such a diagram, each supported by a set of grid lines to support visual interpretation of the data, and each plot representing a different data-ink ratio.

The data-ink ratio increases from left to right in the three versions of the same plot.

Plot A has a high data-ink ratio as most of the ink is dedicated to presenting the data apart from the axis lines, labels, and tick marks necessary to understand the data.

Plot B has a slightly lower data-ink ratio as more ink has been added to draw grid lines. Are the grid lines helpful in understanding the data?

Plot C has a very low data-ink ratio as most of the ink is used for drawing (unhelpful?) grid lines.

Adding more grid lines will reduce the data-ink ratio (increase the amount of non-data ink) which, in principle, is not desirable. One notable exception where grid lines are very helpful is when designing figures intended to be used for manual extraction of data. For example, the Moody diagram shown below relates the friction coefficient in a pipe to the Reynolds number and surface roughness of the pipe. In such cases the addition of grid lines might be very helpful to assist extraction of numbers from the graph. However, for cases where the primary aim of the figure is to illustrate a relationship without necessarily requiring interaction with the figure, the need for grid lines might be less clear. In any case, adding an excessive amount of grid lines (Plot C) compromise the usefulness of the graph for the purpose of data extraction.

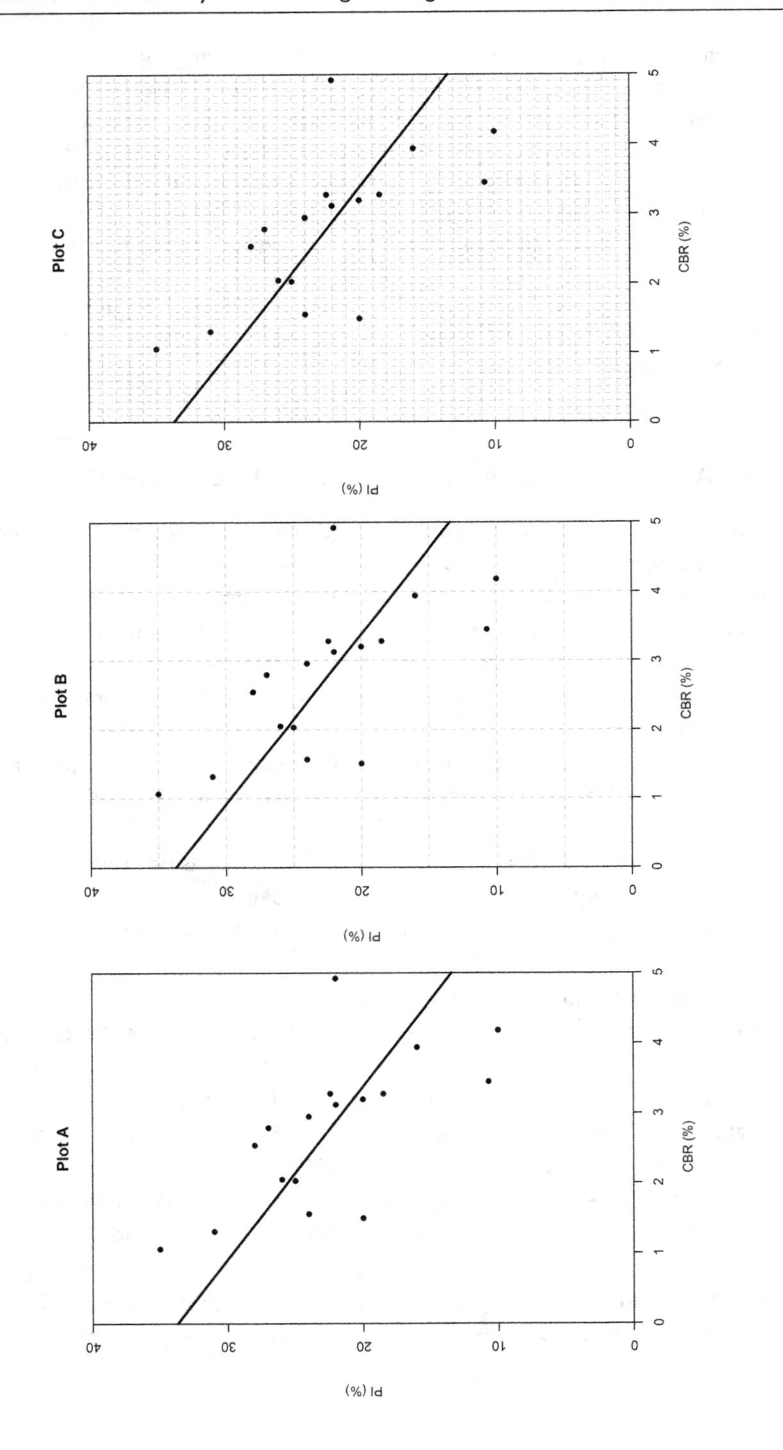

Plot A
Plot B
Plot C
PI (%)
CBR (%)
0
1
2
3
4
5
10
20
30
40

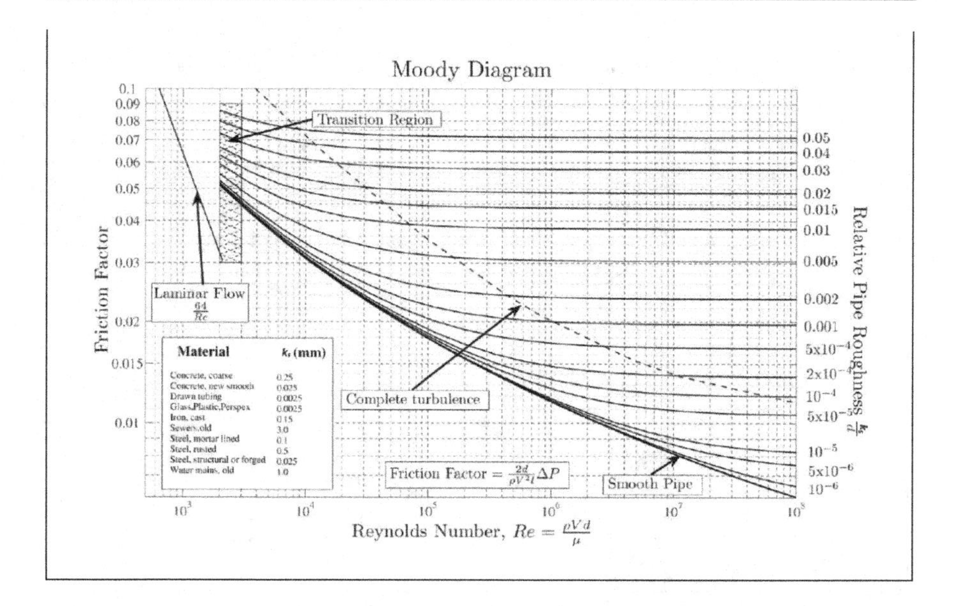

EXAMPLE 4.8 FULL OR HALF FRAMING?

Some graphical designers advocate a reduction in non-data ink by using half frames rather than full frames as shown in the figure below.

Both plots are correct, and it is a matter mostly of personal taste, but as no information regarding the data are lost by using half frames, this option does increase the data-ink ratio.

However, in your drive to increase the data-ink ratio, be careful not to remove essential graph components. It is important to always label the axis appropriately, clearly stating what the numerical values represent and the accompanying physical units as an absolute minimum.

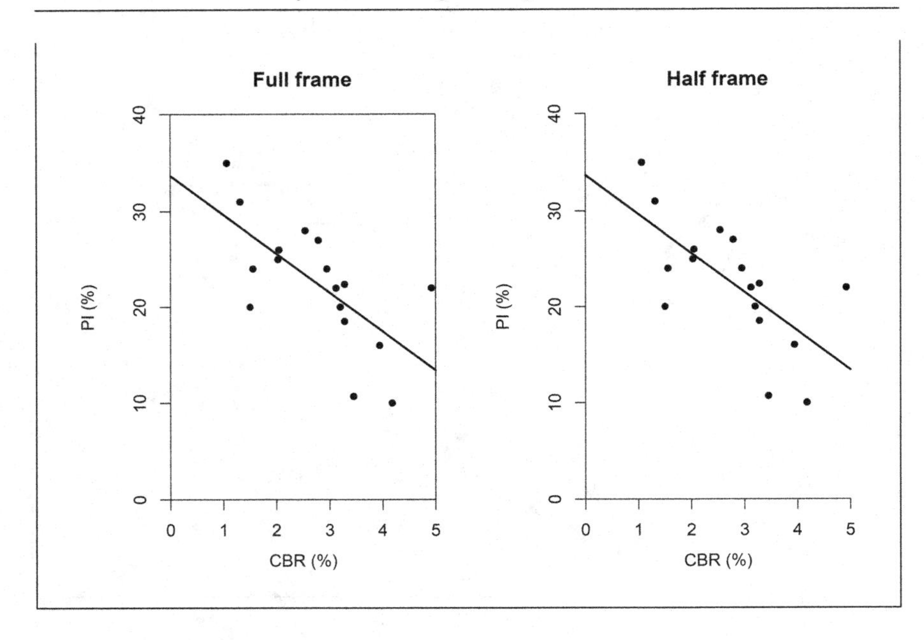

REFERENCE

Tufte, E. R. (2001). *The visual display of quantitative information*. Cheshire, CT: Graphics Press.

Chapter 5

Uncertainty Analysis of Engineering Systems

5.1 INTRODUCTION TO UNCERTAINTY ANALYSIS

Often engineers are faced with complex technical systems where one or more factors are unknown and therefore best represented by a random variable. For example, the performance of a transport system might depend on both the weather conditions and the number of users on any particular date; both of which are unknown and vary randomly from day to day, or even second to second. As a result, the performance of the transport system (e.g. measured as the travel time) is itself a random variable. Another example is structural design where both the applied loads and the material strength are uncertain (i.e. may vary randomly) and, therefore, there is the probability of system failing. The generic situation is sketched in Figure 5.1, where two random variables X_1 and X_2 are input into a system where a mathematical model M transforms the inputs into an output variable named Z. The mathematical model might be a set of equations, but as both input variables are random variables, the output also becomes a random variable. This can be interpreted as the uncertainty in the input parameters being propagated through the system and manifesting itself as uncertainty in the output.

Depending on the complexity of the system in terms of number of input variables and whether the model M is a linear or non-linear model, the mean and variance of the output Z can be expressed as function of the mean and variance of the input variables. Finally, the probability of failure is calculated. In this chapter analytical methods will be introduced which are suitable for relatively simple systems characterised by one or two random input variables.

5.2 FUNCTIONS OF RANDOM VARIABLES, LINEAR SYSTEMS

Consider a simple system where a random variable Z is a *linear* function of n random variables $X_1, \dots X_n$, each characterised by a mean value $\mu_1, \dots, \mu_n$ and variance $\sigma_1^2, \dots, \sigma_n^2$:

$$Z = a_0 + a_1 X_1 + \dots + a_n X_n \tag{5.1}$$

DOI: 10.1201/9781032700373-5

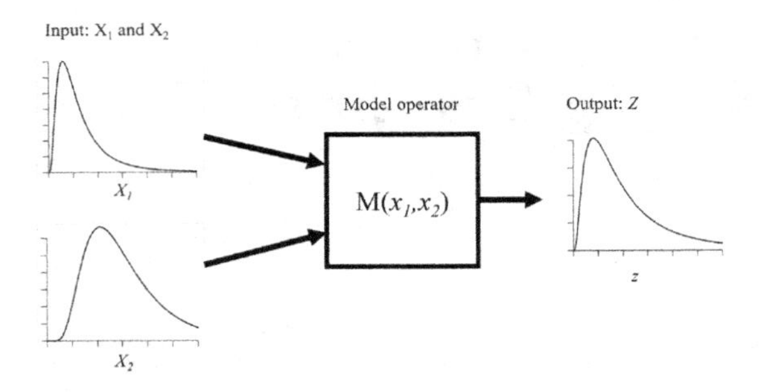

Figure 5.1 Propagation of uncertainty from input variables through the model *M* to output.

where $a_0, a_1, \ldots, a_n$ are constants. This is a linear system, and direct application of the mean operator E gives the following:

$$\begin{aligned} E(Z) &= E(a_0 + a_1 X_1 + \ldots + a_n X_n) \\ &= a_0 + a_1\ E(X_1) + \ldots + a_n\ E(X_n) \\ &= a_0 + \sum_{i=1}^{n} a_i\ \mu_i \end{aligned} \tag{5.2}$$

Next, the variance of the output for a linear system can be determined using the variance operator V as follows:

$$\begin{aligned} V(Z) &= V(a_0 + a_1\ X_1 + \ldots + a_n\ X_n) \\ &= a_1^2\ V(X_1) + \ldots + a_n^2 V(X_n) + 2\sum_{i=1}^{n}\sum_{j=i+1}^{n} a_i\ a_j\ Cov(X_i, X_j) \end{aligned} \tag{5.3}$$

Note here the contribution of the covariance term at the end of Eq. 5.3. If there is a positive covariance between two of the random input variables, then the variance of the output Z is inflated. If the random input variables are all independent, then the covariance is zero.

Equations 5.2 and 5.3 are valid regardless of the distribution of the random input variables $X_1, \ldots, X_n$. It is generally challenging to determine the exact probability distribution of the output Z. A notable exception is for cases where $X_1, \ldots, X_n$ are normally distributed, which leads to Z being itself normally distributed with mean and variance given by Eqs. 5.2 and 5.3, respectively.

Note that a special case of Eq. 5.3 includes the following:

$$V(a_0) = 0 \tag{5.4}$$

meaning that the variance of a constant is zero. Also, constants are free to move out of the variance operator but must be squared when doing so, i.e.:

$$V(a_1X) = a_1^2 V(X) \tag{5.5}$$

For the special case where Z is a linear function of two random variables X_1 and X_2 with the following form:

$$Z = a_0 + a_1\ X_1 + a_2\ X_2 \tag{5.6}$$

the mean and variance of the output Z in Eq. 5.6 are defined from Eqs. 5.2 and 5.3 as:

$$E(Z) = E(a_0 + a_1X_1 + a_2X_2) = a_0 + a_1E(X_1) + a_2E(X_2) \tag{5.7}$$

$$\begin{aligned} V(Z) &= V(a_0 + a_1X_1 + a_2X_2) \\ &= a_1^2 V(X_1) + a_2^2 V(X_2) + 2a_1a_2 Cov(X_1, X_2) \end{aligned} \tag{5.8}$$

For cases where the two random variables X_1 and X_2 are independent, the variance in Eq. 5.8 reduces to $V(Z) = a_1^2 V(X_1) + a_2^2 V(X_2)$.

EXAMPLE 5.1 RELIABILITY OF A STRUCTURAL ELEMENT

Define *reliability* as the probability of not observing a failure.

Consider a cantilever beam of length L in the figure below. The beam is anchored in a wall with a resistance moment capacity M_R. Consider two independent loads: (i) a concentrated load P located at distance L from the wall, and a distributed load w along the length of the beam. What is the reliability of this system?

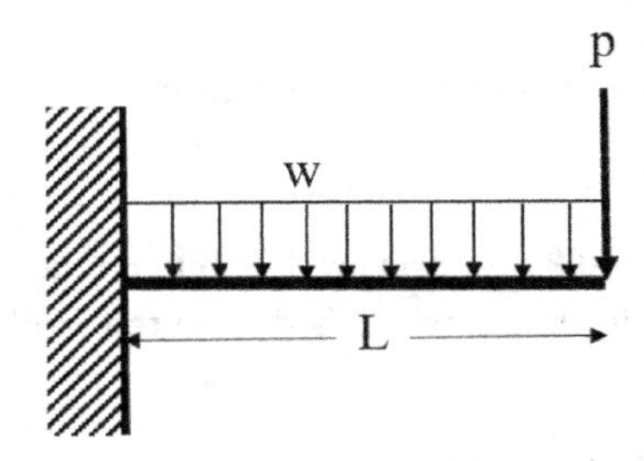

Cantilever example.

Assume that the moment capacity (M_R) and the two loads (p and w) are independent normally distributed random variables with the parameters listed in the following table.

Input Parameters for Cantilever Example

Variable	*Mean Value*	*Standard Deviation*
Length, L	$10\,m$	Not random
Load, w	$\mu_w = 10\,kN/m$	$\sigma_w = 1 kN/m$
Load, P	$\mu_P = 25\,kN$	$\sigma_P = 5\,kN$
Moment, M_R	$\mu_R = 800\,kNm$	$\sigma_R = 150\,kNm$

Note how the standard deviations in the last column of the table have the same units as the mean values of the random variables. Also, the length of the beam is known and, therefore, is not a random variable. In that case, we define the length as the mean value and set the standard deviation to 0 (no variation).

Solution: The performance of the system is measured as the difference between the moment capacity M_R where the beam is anchored into the wall, and the total moment resulting from the concentrated and distributed loads, M_P and M_w, as formulated below:

$$Z = M_R - (M_P + M_w) = M_R - M_P - M_w$$

The system is in a non-failure state if $Z > 0$ and in a failure state if $Z <= 0$. We can therefore define the reliability of the system as the probability of non-failure, as:

$$\text{Reliability } P(Z > 0) = 1 - P(Z < 0)$$

Before this probability can be evaluated, the mean and variance of Z needs to be derived.

The moment due to the point load, M_P, is given as:

$$M_P = PL$$

where L is a constant and P is a normally distributed random variable as defined above. As a result, M_P itself is a normally distributed random variable with the following mean and variance:

$$\mu_{M_P} = E(M_P) = E(PL) = LE(P) = L\mu_P$$

$$\sigma_{M_P}^2 = V(PL) = L^2V(P) = L^2\sigma_P^2$$

Next, the moment from the distributed load is calculated as:

$$M_w = w\frac{L^2}{2}$$

where again L is constant and w is a normally distributed random variable. Consequently, M_w is a normally distributed random variable with mean and variance given as:

$$\mu_{M_w} = E\left(w\frac{L^2}{2}\right) = \frac{L^2}{2}E(w) = \frac{L^2}{2}\mu_w$$

$$\sigma_{M_w}^2 = V\left(w\frac{L^2}{2}\right) = \frac{L^4}{4}V(w) = \frac{L^4}{4}\sigma_w^2$$

Using Eqs. 5.2 and 5.3, the mean and variance of Z can now be calculated as follows:

$$\mu_Z = \mu_{M_R} - \mu_{M_P} - \mu_{M_w} = \mu_{M_R} - L\mu_P - \frac{L^2}{2}\mu_w$$

$$\sigma_Z^2 = \sigma_{M_R}^2 + \sigma_{M_P}^2 + \sigma_{M_w}^2 = \sigma_{M_R}^2 + L^2\sigma_P^2 + \frac{L^4}{4}\sigma_w^2$$

As the system performance Z is a normally distributed random variable, the probability of a failure $P(Z <= 0)$ can be calculated as:

$$P(Z \leq 0) = \Phi\left(\frac{0-\mu_Z}{\sigma_Z}\right) = \Phi\left(\frac{0-\left(\mu_{M_R} - L\mu_P - \frac{L^2}{2}\mu_w\right)}{\sqrt{\sigma_{M_R}^2 + L^2\sigma_P^2 + \frac{L^4}{4}\sigma_w^2}}\right)$$

By inserting the values for mean and variance, the following expression is obtained:

$$P(Z < 0) = \Phi\left(\frac{0-(800-(10)(25)-(10^2/2)(10))}{\sqrt{150^2 + (10^2)(5^2) + (10^4/4)(1^2)}}\right) = \Phi(-0.3015) = 0.38$$

Note that as $z = -0.3015 < 0$ it is necessary to use the relationship $\Phi(z) = 1 - \Phi(-z)$ to use the table in Chapter 2, Appendix A or use the *NORMDIST* function in EXCEL.

EXAMPLE 5.2 RELIABILITY OF A WATER SUPPLY SYSTEM

A community is abstracting water from a river for domestic, agricultural, and industrial consumption. The annual water demand is fixed at 150×10^6 m^3/year. The river is formed by joining of two upstream tributaries. What is the probability of the water availability in any one year being unable to meet the specified water demand?

The annual runoff volume from each tributary, X_1 and X_2, is normally distributed with the mean and variance (in 10^6 m³/year) specified as:

$$X_1 \sim N\left(100, 10^2\right)$$

$$X_2 \sim N\left(80, 8^2\right)$$

Because the flow in the rivers is partly a result of rainfall from the same large-scale weather systems, the annual volumes in the two rivers are positively correlated with a correlation coefficient of $\rho = 0.75$. This means that in dry years, both tributaries are expected to provide less water, and in wet years, an abundance of water will be available from both tributaries.

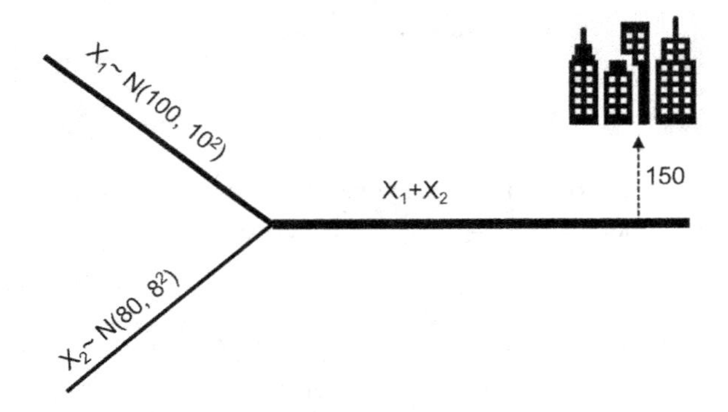

Schematic representation of river and water supply system.

Let $Z = X_1 + X_2$ denote the total annual water availability. Failure to satisfy water demand occurs in years where $Z = X_1 + X_2 < 150$. As both X_1 and X_2 are normally distributed, the sum $Z = X_1 + X_2$ is also normally distributed with mean and variance calculated using Eqs. 5.2 and 5.3:

$$E(Z) = E(X_1) + E(X_2) = 100 + 80 = 180$$

$$V(Z) = V(X_1) + V(X_2) + 2\sigma_1 \sigma_2 \rho = 10^2 + 8^2 + 2 \cdot 10 \cdot 8 \cdot 0.75 = 16.85^2$$

Therefore, the distribution of annual runoff volume in the main river is a normal distribution specified as:

$$Z \sim N\left(180, 16.85^2\right)$$

The probability of not satisfying the water demand can now be calculated as follows:

$$P(Z \leq 150) = P\left(\frac{Z-180}{16.85} \leq \frac{150-180}{16.85}\right) = \Phi(-1.78) = 0.038$$

This means there is a 3.8% probability each year for the water demand not being met by the river system. If this is considered unacceptably high, the community must either (1) increase supply by building a reservoir or develop an alternative water supply source, or (2) lower the demand by saving water, or meeting both options.

Note that if the correlation between annual runoff from the two tributaries had not been considered, consequently the variance of $Z = X_1 + X_2$ would have been $10^2 + 8^2 = 12.8^2$ and the resulting probability of not satisfying the water demand would become:

$$P(X_1 + X_2 \leq 150) = P\left(\frac{X_1 + X_2 - 180}{12.8} \leq \frac{150-180}{12.8}\right) = \Phi(-2.34) = 0.010$$

Ignoring correlation in this case means ignoring the tendency of the two tributaries to supply less water in dry years. It is important to note that by not considering this correlation, the community would have been exposed to a false sense of security by providing a lower probability of failure.

5.3 FUNCTIONS OF RANDOM VARIABLES, NON-LINEAR SYSTEMS

This section introduces an approximate method that allows the mean and variance to be calculated for non-linear systems. This method is based on Taylor approximations (or expansions) and is sometimes referred to as the delta method.

5.3.1 Taylor Approximations

Consider the general case where a random variable Z is a non-linear function of n independent random variables $X_1, \ldots, X_n$, as follows:

$$Z = g(X_1, \ldots, X_n) \tag{5.9}$$

where g is a known non-linear and differentiable function. As the mean and variance operators defined in Eqs. 5.2 and 5.3 only apply to linear systems, it is necessary to first *linearise* the function g using Taylor's approximation.

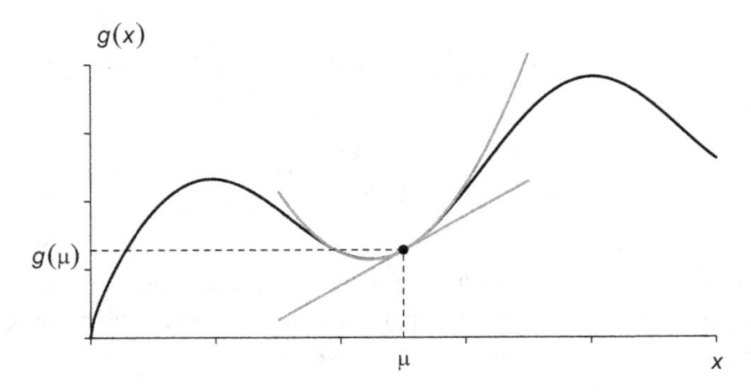

Figure 5.2 Taylor approximation to function $g(x)$ around the point μ. The grey curved line represents second-order Taylor approximation.

The principle behind a Taylor approximation is illustrated in Figure 5.2 and is based on fitting a linear function to approximate values of g in the vicinity of a particular point (such as $x = \mu$). First, consider the case where g is a function of a single input variable x, i.e. $g(x)$. If $g(x)$ can be differentiated in the point $x = \mu$, consequently $g(\mu)$ has a tangent in the point $(\mu, g(\mu))$ defined as follows:

$$y = g(\mu) + g'(\mu)(x - \mu) \tag{5.10}$$

The tangent can be considered a first-order approximation to the non-linear function $g(x)$ which is "reasonably close" for values of x in the close vicinity to the point $x = \mu$. If the first-order approximation is not sufficiently accurate, then a second-order approximation can be developed as follows:

$$y_2 = g(\mu) + g'(\mu)(x - \mu) + \frac{1}{2} g''(\mu)(x - \mu)^2 \tag{5.11}$$

As can be seen on Figure 5.2, the second-order approximation aligns closer with $g(x)$ but it is analytically more complex as a second order term has been added (compare Eqs. 5.10 and 5.11).

EXAMPLE 5.3 DERIVE A TAYLOR APPROXIMATION

Derive the first-order Taylor approximation to the function: $g(x) = \sin(x) + \sqrt{x}$.

First, differentiate $g(x)$ to obtain $g'(x)$ as follows:

$$g'(x) = \cos(x) + \frac{1}{2\sqrt{x}}$$

The first-order Taylor approximation in the point $x = \mu$ is now given as:

$$y_1 = \underbrace{\sin(\mu) + \sqrt{\mu}}_{g(\mu)} + \underbrace{\left[\cos(\mu) + \frac{1}{2\sqrt{\mu}}\right]}_{g'(\mu)}(x - \mu)$$

As all the terms evaluated at the point $x = \mu$ are constant, this equation is a simple straight line with intercept $\sin(\mu) + \sqrt{\mu} - \mu\cos(\mu) - \frac{\mu}{2\sqrt{\mu}}$ and slope $\cos(\mu) + \frac{1}{2\sqrt{\mu}}$.

5.4 UNIVARIATE NON-LINEAR SYSTEMS

Consider the case where the random variable Z is a function of a single random variable X as $Z = g(X)$ where X has a mean value μ_x and variance σ_x^2 In this case, the first-order Taylor approximation to Z around the mean value of the input variable μ_x is given as:

$$Z(X) \approx g(\mu_x) + g'(\mu_x)(X - \mu_x) \tag{5.12}$$

The subscript x is added in Eq. 5.12 to highlight properties related to the input variable. Equation 5.12 shows that Z is approximated as a linear function of X rather than the original non-linear function $g(x)$. Using the mean and variance operators from Eqs. 5.2 and 5.3 on the linearised function gives the following:

$$E(Z) \approx g(\mu_x) \quad \text{as} \quad E((X - \mu)) = 0 \tag{5.13}$$

$$V(Z) \approx [g'(\mu_x)]^2 \sigma_x^2 \tag{5.14}$$

However, as generally $E(g(x)) \neq g(E(X))$, it is sometimes necessary to consider the second-order Taylor expansion to evaluate the mean. This means:

$$Z = g(\mu_x) + g'(\mu_x)(X - \mu_x) + \frac{1}{2}g''(\mu_x)(X - \mu_x)^2 \tag{5.15}$$

Using the mean operator on the right-hand side of the second-order Taylor expansion gives:

$$E(Z) = g(\mu_x) + \frac{1}{2}g''(\mu_x)\sigma_x^2 \tag{5.16}$$

as $E\left((X - \mu_x)^2\right) = V(X) = \sigma_x^2$.

5.5 MULTIVARIATE NON-LINEAR SYSTEMS

The Taylor approximation of the more general case where Z is a function of two (or more) random variables X and Y as $Z = g(X,Y)$ is a straightforward extension of the univariate case. The first-order Taylor approximation of the function g defined by the point $\mu = (\mu_x, \mu_y)$ has the following form:

$$Z(X,Y) = g(\mu) + \frac{\partial g(\mu)}{dx}(X - \mu_x) + \frac{\partial g(\mu)}{dy}(Y - \mu_y) \tag{5.17}$$

where the derivatives are evaluated in the point $\mu = (\mu_x, \mu_y)$. As both $g(\mu_x, \mu_y)$ and the two derivatives are constants, the expression in Eq. 5.17 is a sum of the two random variables X and Y, i.e., a linear system, and the variance of Z therefore is given via Eq. 5.3 as the variance of a sum of random variables:

$$V(Z) = \left[\frac{\partial g(\mu)}{dx}\right]^2 V(X) + \left[\frac{\partial g(\mu)}{dy}\right]^2 V(Y) + 2\left[\frac{\partial g(\mu)}{dx}\right]\left[\frac{\partial g(\mu)}{dy}\right] Cov(X,Y) \tag{5.18}$$

As was the case for the univariate function in Section (5.4), the first-order Taylor approximation to the mean is:

$$E(Z) \approx g(\mu_x, \mu_y) \tag{5.19}$$

If this is not sufficiently accurate, then a second-order Taylor approximation is necessary, as follows:

$$\begin{aligned} Z(X,Y) = g(X,Y) &\approx g(\mu) + \frac{\partial g(\mu)}{\partial x}(X - \mu_x) + \frac{\partial g(\mu)}{dy}(Y - \mu_y) \\ &+ \frac{1}{2}\frac{\partial^2 g(\mu)}{dx^2}(X - \mu_x)^2 + \frac{1}{2}\frac{\partial^2 g(\mu)}{dy^2}(Y - \mu_y)^2 \\ &+ \frac{\partial^2 g(\mu)}{dxdy}(X - \mu_x)(Y - \mu_y) \end{aligned} \tag{5.20}$$

Using the mean operator gives the second-order approximation to the mean of Z as:

$$E(Z) \approx g(\mu) + \frac{1}{2}\frac{\partial^2 g(\mu)}{dx^2}\sigma_x^2 + \frac{1}{2}\frac{\partial^2 g(\mu)}{dy^2}\sigma_y^2 + \frac{\partial^2 g(\mu)}{dxdy} Cov(X,Y) \tag{5.21}$$

Remember that $E\left(\left(X-\mu_x\right)\left(Y-\mu_y\right)\right)=Cov\left(X,Y\right)$ as per Eq. 2.22.

The general principles of the two-dimensional Taylor extension outlined above can be extended to an arbitrarily large number of variables following the same principles.

EXAMPLE 5.4 MEAN AND VARIANCE OF THE FACTOR OF SAFETY

UNIVARIATE

A **system** is designed so that the load, L, is a random variable with mean and variance μ_L and σ_L^2, respectively, but a constant (non-random) resistance, R_0. The factor of safety, F for this system is defined as $F = R_0 / L$. Find the mean and variance, μ_F and σ_F^2 of the factor of safety.

First, define the function: $F = g\left(L\right) = \dfrac{R_0}{L}$

Next, the derivatives required for the first- and the second-order Taylor approximations to $g(L)$ are calculated and evaluated in the mean value μ_L as follows:

$$\frac{dg\left(\mu_L\right)}{dL} = -\frac{R_0}{\mu_L^2}$$

$$\frac{d^2g\left(\mu_L\right)}{dL^2} = \frac{2R_0}{\mu_L^3}$$

Using Eqs. 5.14 and 5.16, the mean and variance of F can now be calculated as:

$$E\left(F\right) \approx \frac{R_0}{\mu_L} + \frac{R_0}{\mu_L^3}\sigma_L^2$$

$$V\left(F\right) \approx \frac{R_0^2}{\mu_L^4}\sigma_L^2$$

MULTIVARIATE

Both resistance, R, and load, L, from the example above are now considered random variables. Resistance has a mean and variance μ_R, σ_R^2, and the covariance between R and L is determined via the correlation coefficient ρ as $Cov(R,L) = \rho\sigma_R\sigma_L$ (see Eq. 2.23 in Chapter 2). Determine the mean and variance of the factor of safety.

First, define a new multivariate function: $F = g(R, L) = \frac{R}{L}$

Next, calculate the derivatives required for the first- and the second-order Taylor approximations and evaluate in the point $\mu = (\mu_R, \mu_L)$ as formulated below:

$$\frac{\partial g(\mu)}{\partial R} = \frac{1}{\mu_L}$$

$$\frac{\partial g(\mu)}{\partial L} = -\frac{\mu_R}{\mu_L^2}$$

$$\frac{\partial^2 g(\mu)}{\partial R^2} = 0$$

$$\frac{\partial^2 g(\mu)}{\partial L^2} = \frac{2\mu_R}{\mu_L^3}$$

$$\frac{\partial^2 g(\mu)}{\partial R \partial L} = -\frac{1}{\mu_L^2}$$

The mean and variance of the factor of safety can now be found using Eqs. 5.21 and 5.18, thus:

$$E(F) \approx \frac{\mu_R}{\mu_L} + \frac{\mu_R}{\mu_L^3}\sigma_L^2 + 0\sigma_R^2 + \left(-\frac{1}{\mu_L^2}\right)\rho\sigma_R\sigma_L$$

$$V(F) \approx \frac{1}{\mu_L^2}\sigma_R^2 + \frac{\mu_R^2}{\mu_L^4}\sigma_L^2 - 2\frac{\mu_R}{\mu_L^3}\rho\sigma_R\sigma_L$$

The complexity of these equations can be reduced considerably if resistance and load are independent random variables, because this will result in $\rho = 0$; therefore, the last term on the right-hand side of both the mean and variance would disappear.

EXAMPLE 5.5 CALCULATE THE VARIANCE OF COMPRESSIVE STRENGTH OF A BRICK WALL

According to Eurocode 6 the characteristic compressive strength of masonry, f_k, is given as:

$$f_k = K f_b^{\alpha} f_m^{\beta}$$

The compressive strength of bricks, f_b, and mortar, f_m, are random variables with mean and variance given as

$$\mu_b = 30, \qquad \sigma_b = 5$$

$$\mu_m = 5, \qquad \sigma_m = 1$$

while assuming the following constants $K = 0.35$, $\alpha = 0.7$ and $\beta = 0.3$ are constants.

To calculate the variance of the combined compressive strength using Eq. 5.18, it is necessary to first consider if the covariance of the strength of bricks and mortar should be considered. However, as there is no physical reason why the strength of the two materials should influence each other, this covariance is assumed equal to zero in this example. To calculate the variance using Eq. 5.18, the function $g(f_b, f_m)$ is defined as:

$$g(f_b, f_m) = f_k = K f_b^{\alpha} f_m^{\beta}$$

Next, the first-order partial derivatives of $g(f_b, f_m)$ are derived for both random variables (strength of bricks and mortar) as:

$$\frac{\partial g}{\partial f_b} = K\alpha f_b^{\alpha-1} f_m^{\beta}$$

$$\frac{\partial g}{\partial f_m} = K f_b^{\alpha} \beta f_m^{\beta-1}$$

Finally, combining all factors into the approximate variance according to Eq. 5.18 gives:

$$V(f_k) \approx \left(K\alpha \mu_b^{\alpha-1} \mu_m^{\beta}\right)^2 \sigma_b^2 + \left(K \mu_b^{\alpha} \beta \mu_m^{\beta-1}\right)^2 \sigma_m^2$$

Note here that the partial derivatives are evaluated in the mean values of the two random variables, i.e. μ_b and μ_m. Inserting numerical values gives the following three terms:

$$V(f_k) \approx 0.5121 + 0.1355 = 0.6476$$

Bricks are responsible for (0.5121/0.6476) × 100% = 79% of the variance, while mortar is responsible for the remaining 21%.

EXAMPLE 5.6 CALCULATE THE PROBABILITY OF EXCEEDING DESIGN FLOOD

A new flood defence is being proposed to defend a city against a design flood with a return period of 100 years. It is proposed that the defence should be constructed with a freeboard of 0.30 m above the water level, h, associated with the design flood. The river channel is rectangular with a width of $b = 10m$ and riverbed slope of $s = 0.01$, and a Manning's number of $n = 1/30$.

The 100-year design flood Q_{100} is a random variable following a normal distribution with a mean value of $200 m^3/s$ and standard deviation $15 m^3/s$, i.e.: $Q_{100} \sim N(200, 15^2)$. What is the probability of the water level of the 100-year flood exceeding the freeboard and inundating the city, assuming the water level is normally distributed?

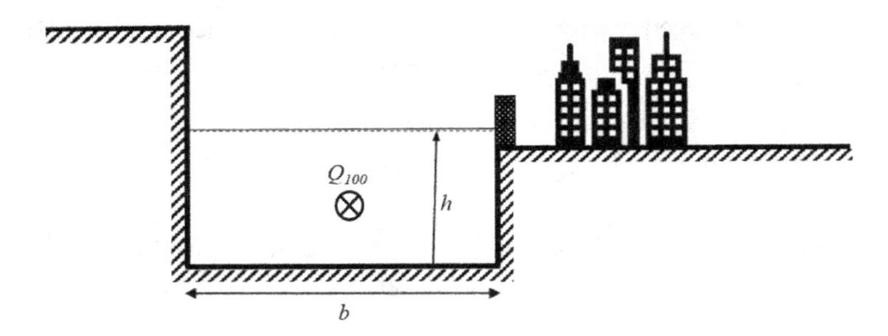

Cross section of river channel, including a flood defence structure, and a community living behind the flood defence.

The first step is to calculate the water level associated with the design flood of $Q_{100} = 200m^3/s$ using Manning's equation relating flow discharge to hydraulic radius, R, river slope, s, cross-sectoral area of flow, A and the Manning's number n as follows:

$$Q = \frac{1}{n} s^{1/2} R^{2/3} A$$

For a rectangular channel, the cross-sectoral area of flow is defined as $A = bh$ and the hydraulic radius as $R = \frac{bh}{b+2h}$, which gives:

$$Q = \frac{1}{n} s^{1/2} \left(\frac{bh}{b+2h} \right)^{2/3} (bh)$$

There is no analytical solution to this equation for the water depth h, but Newton's method can be used to solve this equation, which results in a water depth of $h = 3.94\,m$.

To find the variance of the water level h, the non-linear Manning's equation is linearised using Taylor's equation, assuming only flow Q and h are random variables:

$$Q \approx Q(h_0) + \frac{dQ}{dh}(h - h_0)$$

Using the variance operator, a relationship between the variance of the design flood and the variance of the water level is obtained as follows:

$$V(Q) = \left(\frac{dQ}{dh}\right)^2 V(h)$$

The derivative $\frac{dQ}{dh}$ is derived by differentiating the Manning's equation as:

$$\begin{aligned}\frac{dQ}{dh} &= \frac{1}{n}S^{1/2}\left(\frac{2}{3}R_h^{-1/3}\frac{dR_h}{dh} + R_h^{2/3}\frac{dA}{dh}\right)\\ &= Q\left(\frac{2}{3}\frac{1}{R_h}\frac{dR_h}{dh} + \frac{1}{A}\frac{dA}{dh}\right)\end{aligned}$$

For a rectangular channel, the area is $A = bh$ and therefore:

$$\frac{dA}{dh} = b$$

and the hydraulic radius is: $R_h = \frac{bh}{2h+b}$. Therefore, its derivative relative to h becomes:

$$\frac{dR_h}{dh} = \frac{b^2}{(2h+b)^2}$$

By combining the above three equations and reorganising them, the following expression is obtained:

$$\frac{dQ}{dh} = Q\left(\frac{5b+6h}{3h(2h+b)}\right) = 200m^3/s\left(\frac{5\times 10m + 6\times 3.94m}{3\times 3.94m\times(2\times 3.94m + 10m)}\right) = 69.74m^2/s$$

The variance of the water level h can now be calculated by inserting the values for $b = 10m$, $h = 3.94m$, $Q_{100} = 200m^3/s$, and $V(Q_{100}) = 15^2$ as:

$$V(h) = \left(\frac{dQ}{dh}\right)^{-2} V(Q_{100}) = (69.74m^2/s)^{-2} \times (15m^3/s)^2 = (0.215m)^2$$

Assuming water level is a normal distribution as $N \sim N(3.94, 0.27^2)$, the probability of exceeding the 0.30 m additional freeboard can be estimated as:

$$P(H > (h + 0.30m)) = P\left(\frac{H - 3.94}{0.22} > \frac{3.94 + 0.30 - 3.94}{0.22}\right)$$
$$= 1 - \Phi(1.39) = 0.082$$

This probability of failure is a result of uncertainty in specifying the design event itself. There is additional failure risk from experiencing a flood with a return period exceeding the design event (e.g. an event with a return period of 200 years). Therefore, the total risk faced by the community could be greater than the risk calculated here.

Chapter 6

Monte Carlo Simulations

6.1 COMPUTER-BASED ESTIMATION OF UNCERTAINTY

Often engineers need to analyse the failure probability of more complex systems for which no simple analytical solution is available, and where one or more factors are uncertain. In such cases, the methods outlined in Chapter 5 will often not be practical. Instead, numerical methods based on computer-generated realisations of random variables can be used to investigate the uncertainty and failure probability. These methods are collectively known as Monte Carlo simulation techniques, with Monte Carlo referring to the central position of randomness and probability in gambling.

Consider a complex engineered system where a model operator M can transform a set of input values X to an output value Z as illustrated in Figure 6.1.

This model M is in principle similar to the multivariate models considered in section 5.3, the main difference being that M now represents a more complex set of model equations rather than a single analytical expression (linear or non-linear). The model, M, can be as complex and non-linear as required; the only constraint is computing power to carry out 1000s of simulations.

If the input values X as well as the parameters in the model M are all known with absolute certainty (i.e. not random variation), then the output Z is determined with absolute certainty, i.e. same input result in same output every time the model is run. However, if one or more of the input variables is classified as a random variable, then the uncertainty in X will propagate through the system via the model M and manifest itself as uncertainty of the output Z. In other words, if X is a random variable, then Z is also a random variable. The output Z can be summarised to reflect the anticipated performance of the system under randomly varying input conditions.

DOI: 10.1201/9781032700373-6

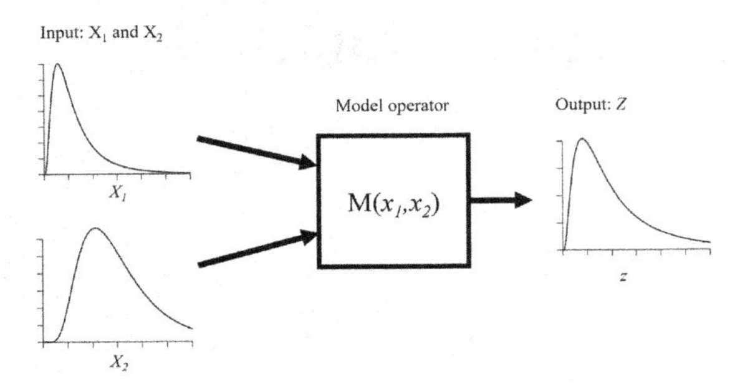

Figure 6.1 Model operator M transforming input in the form of random variables X_1 and X_2 to output Z, also a random variable.

6.2 GENERATION OF REALISATIONS OF RANDOM VARIABLES

Practical use of the Monte Carlo method relies on the ability to generate realisations of the random variables representing the input. Consider that an input variable is a random variable X described by a probability distribution with a known cumulative distribution function (cdf) $F(x)$. A realisation of the random variable X can now be generated by first using a computer to generate a random realisation from a uniform distribution bounded between 0 and 1. Most numerical software tools used for data analysis will have an in-built function that allows such realisation to be generated. For example, in EXCEL the function *RAND*( ) or *random*( ) in Python will generate a random number between 0 and 1. Denote the numerical value of this random realisation as u.

Next, the generated value of u is interpreted as a realisation of the value of the cdf for the chosen distribution, which is naturally bounded between 0 and 1.

Finally, a realisation, x_u, of the random variable X is defined as the uth quantile of the distribution and therefore obtained by solving the equation:

$$F(x_u) = u \rightarrow x_u = F^{-1}(u), \qquad u \in [0,1] \tag{6.1}$$

The value of x_u can now be used as input into the model M and the assigned output Z recorded.

EXAMPLE 6.1 GENERATE RANDOM ELECTRICITY DEMAND DATA

Annual electricity demand within a particular building is a random variable X described by a normal distribution with mean value $\mu = 38.1KW/\left(m^2 \times year\right)$ and standard deviation $\sigma = 17.95KW/\left(m^2 \times year\right)$. To generate a random realisation from this distribution, first generate a random variate from a uniform distribution u defined on the interval $[0,1]$ and then use Eq. 6.1 to transform this value into an random realisation from the normal distribution with mean α and variance σ^2 as defined above. The sequence is illustrated in the figures below, starting by generating $N = 10,000$ random numbers between 0 and 1 from a uniform distribution. A summary of these $N = 10,000$ realisations is shown in the histogram (top) showing a uniform distribution of values across the interval from 0 to 1. Next, each of the values u is transformed into an equivalent realisation of the annual electricity demand using Eq. 6.1 (middle). Finally, the $N = 10,000$ realisations of the annual electricity demand are summarised as a histogram (bottom) which has the exact shape of a normal distribution with a mean value and standard deviation as specified.

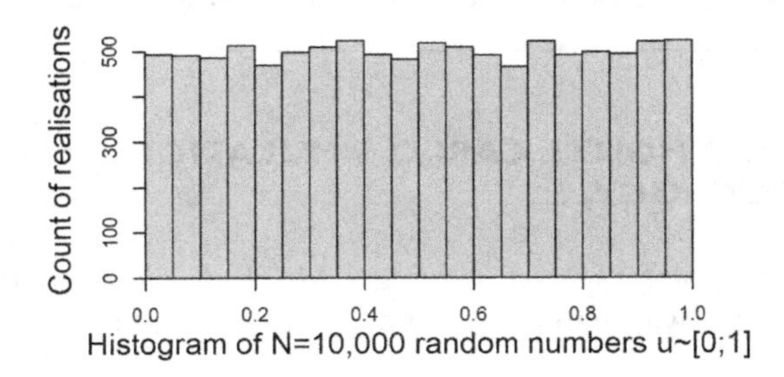

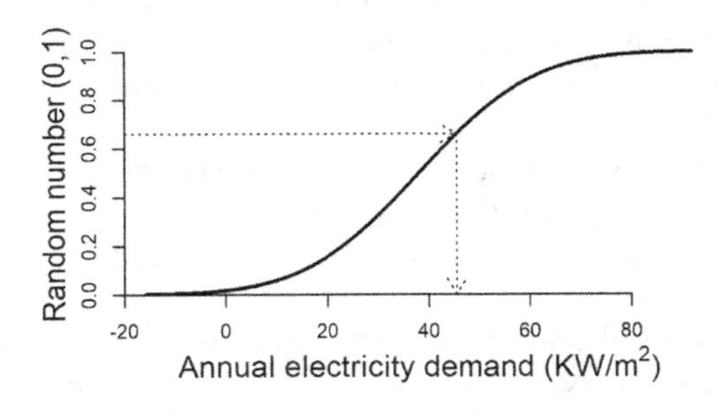

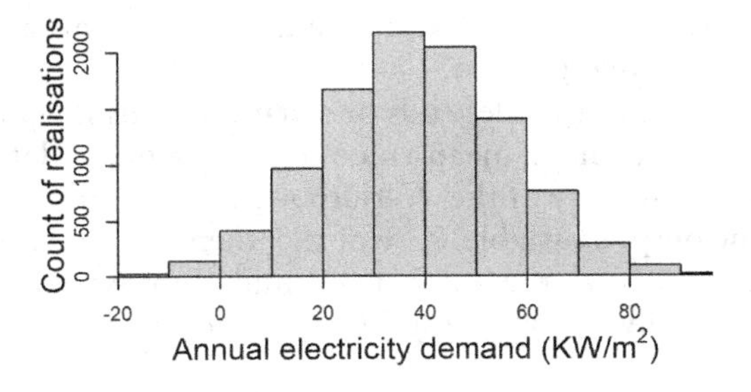

Translation of uniform to normally distributed random realisations

Note that in this case, it is possible to draw realisations that are negative, which is meaningless when discussing electricity use (unless this refers to generating energy within the building).

In practice, if the probability of generating a negative value is negligible, then the negative value can be discarded and a new value generated. In this case the probability of generating a negative annual electricity demand is:

$$P(X < 0) = \Phi\left(\frac{0 - 38.1}{17.95}\right) = 0.0169$$

Thus, we would expect about 1.7% of the realisations to be less than 0. If the probability of generating negative values is considered too great, then an alternative distribution model should be considered. For example, a model that has a natural lower bound of zero such as log-normal or gamma distribution.

6.3 THE BASIC MONTE CARLO SIMULATION METHODOLOGY

The basic methodology for conducting a Monte Carlo simulation of a system characterised by one or more random input variables is as follows:

1. Specify a distribution for each random input variable.
2. Specify the number of random realisations, N.
3. Generate a random realisation of each input variable, x (see Example 6.1).
4. Use the deterministic model M to generate output, z.
5. Repeat steps 3 and 4 a total of N times.
6. Summarise the N output values of Z as required.

There are no specific rules for how many realisations are needed in step 2, but somewhere between $N = 500$ and $N = 10{,}000$ are often used in practice. It is recommended to check the robustness of the results by running the Monte Carlo simulations with different number of realisations and then check if the output converges.

The summary of the output depends on what the analyst wants to communicate, ranging from a simple mean value of Z to a more elaborate graphical and numerical summary of the N output data. If a failure threshold can be defined for the output variable Z (e.g. z_0), then the probability of failure is estimated as by counting the number of simulations where Z exceeds (or falls below) z_0 and divide by the total number of simulations, N.

EXAMPLE 6.2 PROBABILITY OF FAILURE OF CANTILEVER BEAM

In Example 5.1 an analytical solution for the probability of failure for a cantilever beam was developed, assuming both loads and resistance to be normally distributed and independent. The same probability of failure will now be estimated using Monte Carlo simulations.

The first step is to specify the distribution of each of the random input variables. The assumption of a normal distribution for both the resistance and total loads is maintained from Example 5.1, and for ease of notation the total resistance and load are denoted R and Q respectively:

$$R \sim N\left(\mu_{Mr}, \sigma_{Mr}^2\right)$$

$$Q \sim N\left(L\mu_P + L^2/2\mu_w, L^2\sigma_P^2 + L^4/4\sigma_w^2\right)$$

where the numerical values of the terms are defined in Example 5.1. Next, to calculate the probability of failure of the beam using Monte Carlo simulation the rest of the basic procedure is as follows:

1. Generate a random realisation of the resistance, r.
2. Generate a random realisation of the load, q.
3. If $q > r$, then a failure is recorded for the experiment.
4. Repeat steps 1 to 3 a total of N times.
5. Estimate the probability of failure as the fraction of the N experiments where a failure was recorded. In this example N was set to 10,000.

For each of the $N = 10,000$ experiments the generated pairs of load and resistance are plotted against each other with resistance on the x-axis and load on the y-axis. The 1:1 line is added to the plot so that all pairs located above the line are characterised by the load exceeding the resistance, i.e. a failure and the region below the 1:1 line represents a safe system. The figure also includes depictions of the two probability density functions representing the load (vertical) and resistance (horizontal). Note that most of the simulated points are located in the area in the middle of the plot where both distributions are characterised by the highest probability.

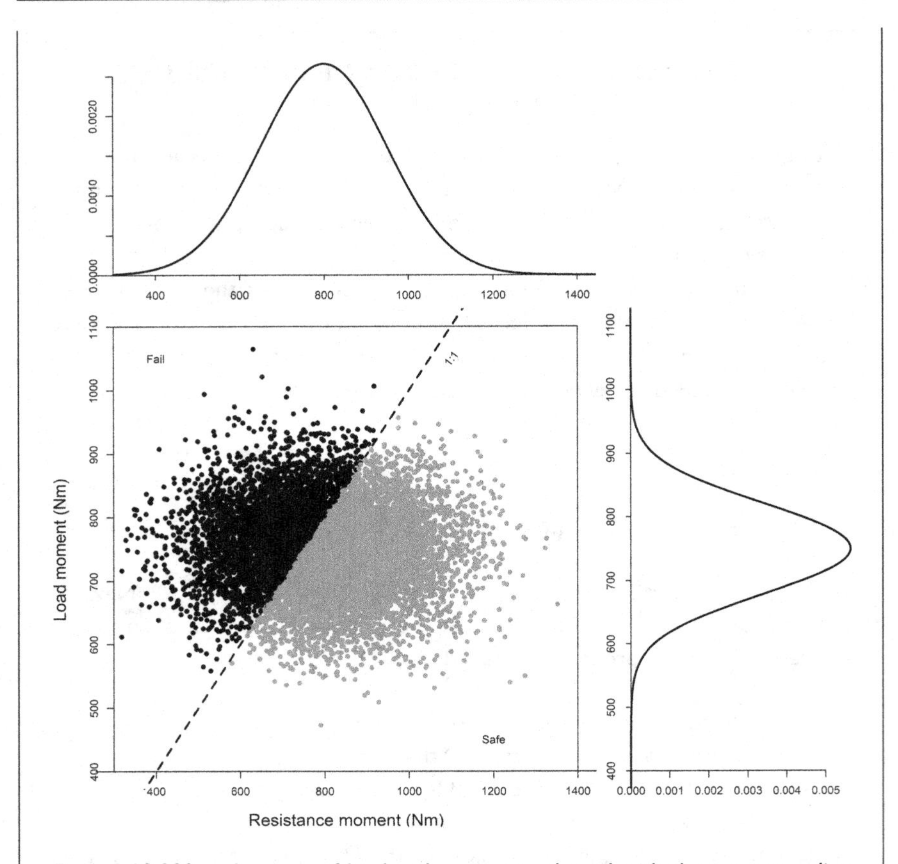

Figure A 10,000 realisations of load and resistance plotted with the corresponding probability distributions.

The number of simulated pairs where load exceeds resistance ($q > r$) equals 3804 (black dots in upper left corner), and therefore the probability of failure of the system can be estimated to be 3804/10000 = 0.3804, which is very close to the analytical answer of 0.38 from Example 5.1.

Using some of the data visualisation techniques discussed in Chapter 4, the output from the Monte Carlo simulations can be summarised in other ways. For example, a histogram showing the N simulated differences in load and resistance. The histogram is made up of 16 bins each with a width of 100 Nm spanning the interval −800 to 800 Nm.

The right-hand side (black) of the histogram represents the generated differences ($Q - R$) where the system would fail (load > resistance). As can be seen, the black area is slightly smaller than the grey area representing the probability of non-failure, which is consistent with the numerical result that the probability of failure is 0.38. In addition, the histogram shows the range and spread of numerical values the difference can possibly take.

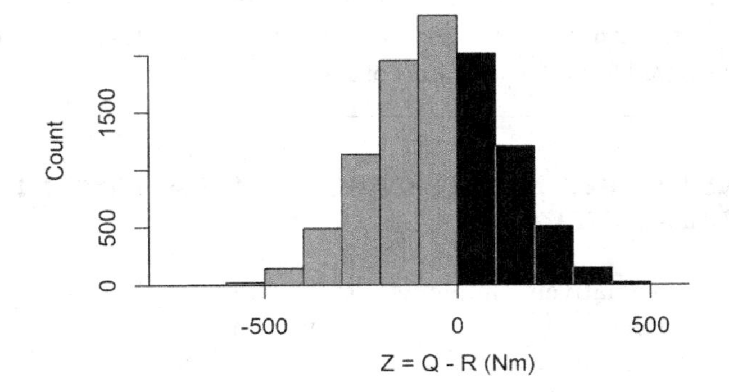

Figure B Histogram of differences between load and resistance.

On both plots, the negative and positive values have been highlighted to emphasise that failure occurs when the load is greater than the resistance, i.e. negative values of $Q - R$ using the colour black. These are only two examples of how to visualise the results, but there are many other graphical ways to summarise datasets, for example using box plots.

It is important that a sufficiently high number of simulations N have been used to adequately cover the expected range of values of load and resistance. Figure C shows how the estimated probability of failure of the beam changes when the number of Monte Carlo simulations increases (black line) from $N = 5$ up to $N = 10^6$ simulations. The graph also compares the simulation results to the analytical solution derived in Example 5.1 (probability of failure = 0.38) shown as the horizontal hatched (grey) line.

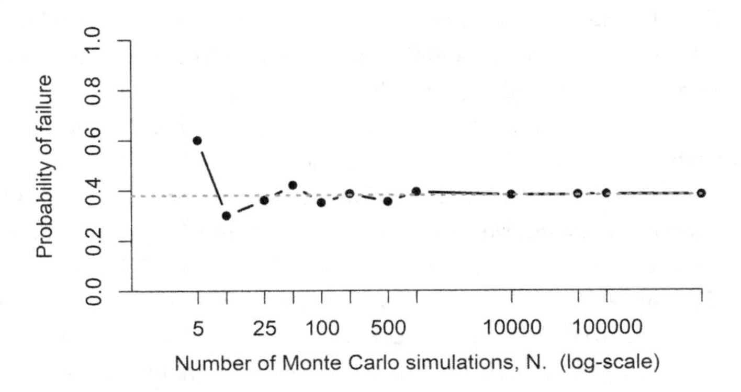

Figure C Convergence of failure probability as the number of Monte Carlo simulations are increased.

When using the output from only a few simulations to estimate the failure probability, it is evident that the resulting estimate is unstable. However, once the number of Monte Carlo simulations increases to $N = 1{,}000$ or more, then the estimated failure probability is seen to convert toward the correct solution derived from the analytical considerations.

6.4 CONDUCTING MONTE CARLO SIMULATIONS IN A SPREADSHEET

Implementing a relatively simple Monte Carlo simulation in a spreadsheet requires a basic level of knowledge of how to use spreadsheets and the in-built functions supporting statistical calculations. Details of readily available in-built EXCEL functions that are helpful for generating random realisations and summarising data are provided in Chapter 14 of this book. Beyond the implementation of the necessary functions and equations, it is also important to pay close attention to the design of the spreadsheet to ensure maximum user-friendliness. The sheer number of calculations required for a Monte Carlo simulation can quickly render the spreadsheet impossible to navigate for other users, or even for the creator when revisiting the calculations after some time has elapsed. Here, Example 6.3 shows one possible implementation of a Monte Carlo simulation in EXCEL, including an example of how to structure the spreadsheet to manage the flow of information, including the input–calculation–output sequence in a user-friendly design.

EXAMPLE 6.3 IMPLEMENT MONTE CARLO IN EXCEL

Create three different worksheets. First, create an "input" sheet where the numerical values of constants and the mean and standard deviation of random variables are characterised. Figure A shows an example of an input sheet used to solve the problem introduced in Example 5.1 (and Example 6.2), considering the probability of failure of the cantilever beam.

Next, create a "calculation" sheet. This sheet is where the actual Monte Carlo simulations themselves are conducted. A Monte Carlo simulation consists of a large number of random realisations (e.g. n=10,000), and in EXCEL each of these realisations will be assigned a row in the "calculation" sheet. For each random variable, create a column where a uniform random number between 0 and 1 is created using the *RAND()* function. Next, create a row where the equivalent random realisation is calculated for a random variable

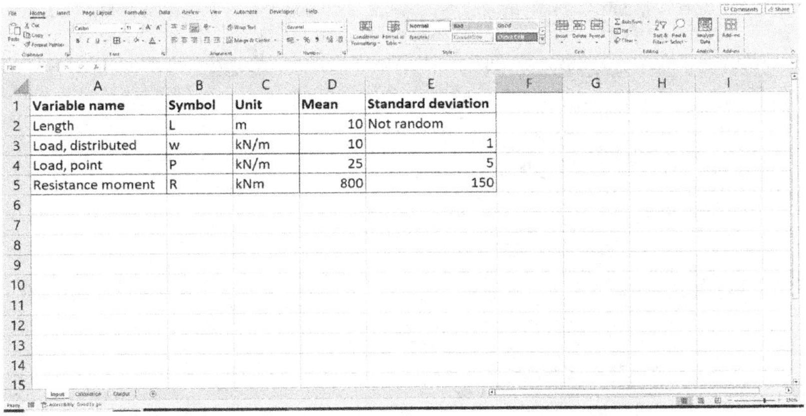

	A	B	C	D	E
1	**Variable name**	**Symbol**	**Unit**	**Mean**	**Standard deviation**
2	Length	L	m	10	Not random
3	Load, distributed	w	kN/m	10	1
4	Load, point	P	kN/m	25	5
5	Resistance moment	R	kNm	800	150

Figure A Design input sheet for Monte Carlo simulations.

by calculating the equivalent quantile as per Eq. 6.1 and using the distribution parameters listed in the "input" sheet.

As discussed in Example 5.1, the difference between the moment generated by the loads and resistance is measured using a random variable Z as:

$$Z = M_R - M_P - M_w = M_R - PL - w\frac{L^2}{2}$$

Therefore, add a final column where values of Z are calculated for each set of realisations. Figure B shows an implementation of the calculations in EXCEL.

In this example, the random realisations of all three variables are based on the normal distribution (as specified in Example 5.1). Thus, the *NORM.INV* function can be used. If other distributions have been specified (e.g. the log-normal distribution), then it is necessary to first estimate the model parameters in the "input" sheet and then find the appropriate version of the EXCEL *INV* function. Details of the different in-built EXCEL functions supporting statistical distributions covered in this book can be found in Chapter 14. Also, remember that it is necessary within equations to fix the reference to cells containing parameter values before you copy and paste equations. For example, to fix the reference to a value in cell D2 (length of beam), change the notation to \$D\$2 in the EXCEL equation.

Sim no.	Rand, w	Rand, P	Rand, R	Load, w	Load, P	Res, R	Z = MR-Mp-Mw
1	0.34757	0.97148	0.92369	9.60812	34.5149	1014.55	188.992
2	0.92414	0.8317	0.85852	11.4335	29.8046	961.051	91.332
3	0.16748	0.21954	0.61946	9.03583	21.1313	845.61	182.505
4	0.4921	0.5604	0.18212	9.98019	25.7599	663.904	-92.7047
5	0.47844	0.90344	0.14386	9.94593	31.5071	640.531	-171.836
6	0.01536	0.5704	0.73764	7.83927	25.887	895.414	=G7-Input!$D$2*Calculation!F7-Calculation!E7*(Input!$D$2^2)/2
7	0.4783	0.17906	0.38159	9.94559	20.4052	754.802	53.4707
8	0.05558	0.73725	0.59078	8.40698	28.1744	834.432	132.339
9	0.2647	0.33127	0.53146	9.37108	22.818	811.841	115.107
10	0.39561	0.64319	0.87963	9.73526	26.835	975.973	220.86
11	0.20635	0.17534	0.47776	9.18084	20.3336	791.633	129.254
12	0.16502	0.7382	0.96303	9.02598	28.189	1068.05	334.858
13	0.56853	0.12691	0.50323	10.1726	19.2943	801.214	99.6392
14	0.27936	0.01003	0.70583	9.41526	13.3747	881.189	276.678
15	0.62174	0.24678	0.11029	10.3101	21.5767	616.25	-115.02
16	0.77593	0.90885	0.86335	10.7585	31.6684	964.324	109.715
17	0.86038	0.30542	0.62972	11.082	22.4556	849.667	71.0097
18	0.537	0.64687	0.01699	10.0929	26.8844	481.963	-291.525
19	0.38714	0.53757	0.87913	9.71321	25.4716	975.6	235.224
20	0.85557	0.86499	0.11287	11.0606	30.5151	618.286	-239.898

Figure B Example of implementation of Monte Carlo simulations in the "calculation" sheet. Notice that references to the model parameter for length L is fixed and use the value in cell D2 in the input sheet. Also, note that this cell reference has been fixed in the equation using the $D$2 reference rather than just D2, allowing copy/paste of the function and its references to all rows.

Finally, the "output" sheet contains the results of the analysis. In the most basic example, this could include, for example, the probability of failure and a histogram of the 10,000 simulated values of the performance function Z. In this case, the beam fails for cases where $Z < 0$. Thus, the probability of failure is approximated as the number of rows in the "calculation" sheet containing a value of Z less than zero, as shown in Figure C. For reference to the *COUINTIF* function, see Chapter 14.

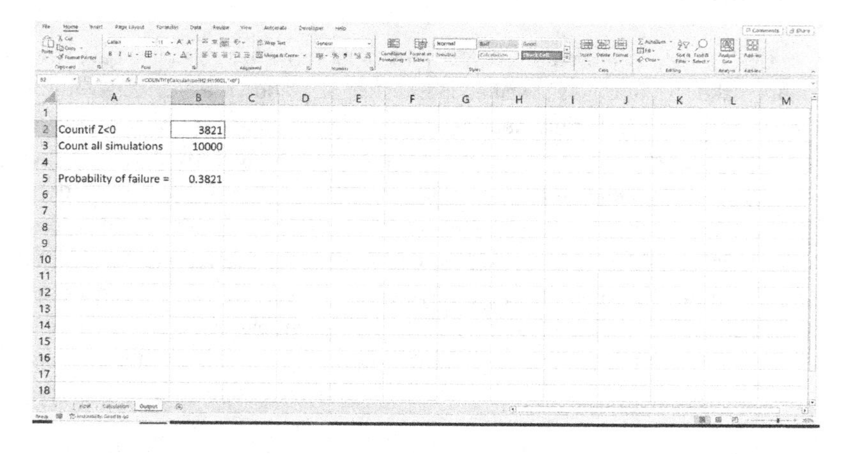

Figure C Geometric definition of the reliability index β as the shortest distance between the origin (0,0) and the border between failure and safety.

Chapter 7

Structural Reliability Analysis

7.1 THE LIMIT STATE FUNCTION

In the context of civil engineering, a failure is broadly defined as a situation where a structure can no longer perform its intended function. The boundary between a failure and a safe state of the structure is defined as the *limit state*, and it is determined mathematically by a *limit state function* (or performance function). For example, consider a system with a resistance (or strength) R and a load L, where both resistance and load are random variables. The system fails when the load exceeds the resistance, and the corresponding limit state function $g(R,L)$ is defined as

$$g(R,L) = R - L \tag{7.1}$$

If $g(R,L) \geq 0$ the structure is safe (or in survival) mode, while values of $g(R,L) < 0$ indicate failure. The corresponding probability of failure p_f is defined as

$$p_f = P(g(R,L) < 0) \tag{7.2}$$

These concepts can be easily generalised to consider structures more generally such that:

$$\text{Safe : Resistance} \geq \text{Load}$$

$$\text{Fail : Resistance} < \text{Load}$$

If the state of a structure is defined according to a number of random state variables $X_1, X_2, \ldots, X_n$, for example different loads and/or aspects contributing to the resistance, then the limit state function is a function (either linear or non-linear) of these variables such that:

$$\begin{aligned} &g(X_1, X_2, \cdots, X_n) > 0, && \text{Safe state} \\ &g(X_1, X_2, \cdots, X_n) = 0 && \text{Boundary between safe and fail} \\ &g(X_1, X_2, \cdots, X_n) \leq 0 && \text{failed state} \end{aligned}$$

DOI: 10.1201/9781032700373-7

7.2 THE RELIABILITY INDEX β

Consider again the limit state function in Eq. 7.1 characterised by two random variables, a resistance R and a load L, as:

$$g(R,L) = R - L$$

Assume R and L are random variables with mean values μ_R and μ_L and variance σ_R^2, σ_L^2. By using the expectation and variance operators introduced in section 5.2 and assuming independence between R and L, the mean and variance of $g(R,L)$ can be calculated as:

$$\mu_g = E(R - L) = \mu_R - \mu_L \tag{7.3}$$

$$\sigma_g^2 = V(R - L) = \sigma_R^2 + \sigma_L^2 \tag{7.4}$$

The subscripts g, R, and L refer to the limit state function, the resistance, and the load.

The *reliability index*, denoted β, is defined as the ratio between the mean and standard deviation of the limit state function (Eqs. 7.3 and 7.4) as:

$$\beta = \frac{\mu_g}{\sigma_g} \tag{7.5}$$

Inserting the expressions for the mean and variance of the function g, derived in Eqs. 7.3 and 7.4, into the definition of the reliability index in Eq. 7.5 gives the following expression for the reliability index:

$$\beta = \frac{\mu_g}{\sigma_g} = \frac{\mu_R - \mu_L}{\sqrt{\sigma_R^2 + \sigma_L^2}} \tag{7.6}$$

Assuming both R and L are independent and normally distributed, then the reliability index can be defined by considering the standardised normal distribution of the difference $R - L$ as:

$$p_f = P(\text{failure}) = P(R - L \le 0) = P\left(Z \le \frac{0 - (\mu_R - \mu_L)}{\sqrt{\sigma_R^2 + \sigma_L^2}}\right) = \Phi(-\beta) \tag{7.7}$$

where Φ is the standard normal distribution $N(0,1)$ tabulated in Chapter 2 (Appendix A) or evaluated using the *NORMDIST* function in EXCEL. The expression in Eq. 7.7 shows a direct link between the reliability index and the probability of failure with selected values shown in Table 7.1.

A geometric interpretation of the reliability index can be developed as follows: assume the system is resting in the mean values of both resistance and

Table 7.1 Reliability Index versus Probability of Failure

Reliability Index, β	*Probability of Failure, p_f*
1.28	10^{-1}
2.33	10^{-2}
3.09	10^{-3}
3.71	10^{-4}
4.26	10^{-5}
4.75	10^{-6}
5.19	10^{-7}
5.62	10^{-8}
5.99	10^{-9}

load, i.e. $R = \mu_R$ and $L = \mu_L$. For a linear system, the reliability index can then be interpreted as the minimum geometrical distance between the system in a state where all variables rest in their (safe) mean values and the failure space as illustrated in Figure 7.1. Note that in Figure 7.1 the random variables load and resistance (L and R) are represented by their normalised values Z_R and Z_L, which both follow a standardised normal distribution with mean 0 and variance of 1. The relationship between R and L and the normalised variable are derived as:

$$
\begin{aligned}
Z_R &= \frac{R-\mu_R}{\sigma_R} \sim N(0,1) \quad \Rightarrow \quad R = \mu_R + Z_R\sigma_R \\
Z_L &= \frac{L-\mu_L}{\sigma_L} \sim N(0,1) \quad \Rightarrow \quad L = \mu_L + Z_L\sigma_L
\end{aligned} \tag{7.8}
$$

Substituting the expressions for resistance and load, R and L, from Eq. 7.8 into the definition of the limit state function in Eq. 7.1 allows the limit state to be expressed in terms of the normalised values as:

$$
g(Z_R, Z_L) = \mu_R + Z_R\sigma_R - \mu_L - Z_L\sigma_L \tag{7.9}
$$

Then consider the border between the failure and the safe regions defined via the limit state function as $g(Z_R, Z_L) = 0$. The geometrical representation of this border can be defined by setting Eq. 7.9 equal to zero and reorganising as:

$$
Z_L = \frac{\mu_R - \mu_L}{\sigma_L} + Z_R\frac{\sigma_R}{\sigma_L} \tag{7.10}
$$

Next, through basic geometrical considerations it can be shown that the minimum distance (as indicated by the dashed line on Figure 7.1) from the

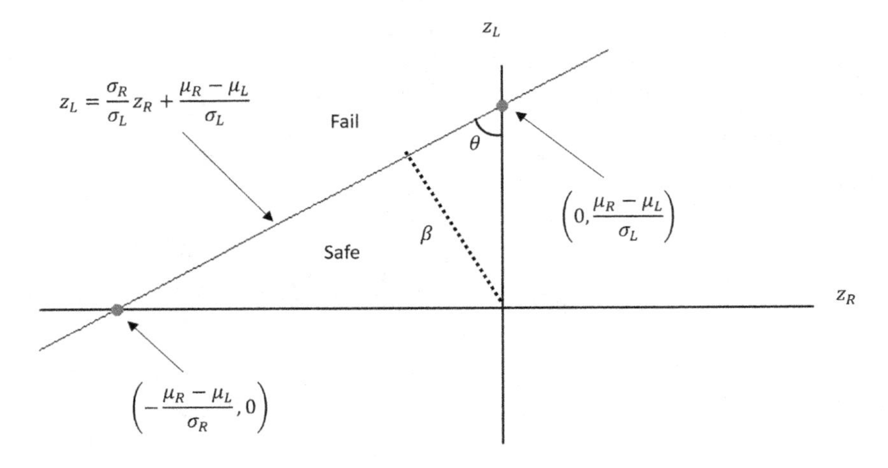

Figure 7.1 Geometric definition of the reliability index β as the shortest distance between the origin (0,0) and the border between failure and safety.

origin $(z_R = 0, z_L = 0)$ to the failure border $g(z_R, z_L) = 0$ is the same as Eq. 7.6 above, i.e.:

$$\beta = \frac{\mu_R - \mu_L}{\sqrt{\sigma_R^2 + \sigma_L^2}} \tag{7.11}$$

Hint: consider the different ways of expressing $\sin(\theta)$ from Figure 7.1.

Finally, the reliability index can be interpreted as representing the number of standard deviations, σ_g, the failure border is located away from the origin which can be realised through a simple reorganisation of Eq. 7.11 as:

$$\mu_R - \mu_L = \beta\sqrt{\sigma_R^2 + \sigma_L^2} \tag{7.12}$$

Thus, the larger the reliability index, the further the failure border is located away from the origin, as also shown on Figure 7.1.

EXAMPLE 7.1 CALCULATE RELIABILITY INDEX OF SIMPLE SYSTEM

A vertical pile is impacted by a load L and has a resistance R; both L and R are normally distributed and independent random variables with the following statistical characteristics:

Load : $\mu_L = 5KN$ and $\sigma_L = 1.5KN$

$$Resistance : \mu_R = 10KN \text{ and } \sigma_L = 2KN$$

This is a simple linear system where the limit state function is defined as:

$$g(R,L) = R - L$$

The corresponding reliability index for this pile is calculated using Eq. 7.6 as:

$$\beta = \frac{10-5}{\sqrt{2^2 + 1.5^2}} = 2.00$$

This result corresponds to a probability of failure of:

$$p_f = \Phi(-\beta) = \Phi(-2.00) \approx 0.0026$$

where the evaluation of the Φ function is done using either the tabulated values in Appendix A of Chapter 2 or the *NORMDIST* function in EXCEL.

The figure below shows the pdf of the limit state function $R-L$. Note how the distance from the failure point to the mean value of the limit state function $\mu_R - \mu_L$ is expressed as β times the standard deviation of the limit state function $\sqrt{\sigma_R^2 + \sigma_L^2}$.

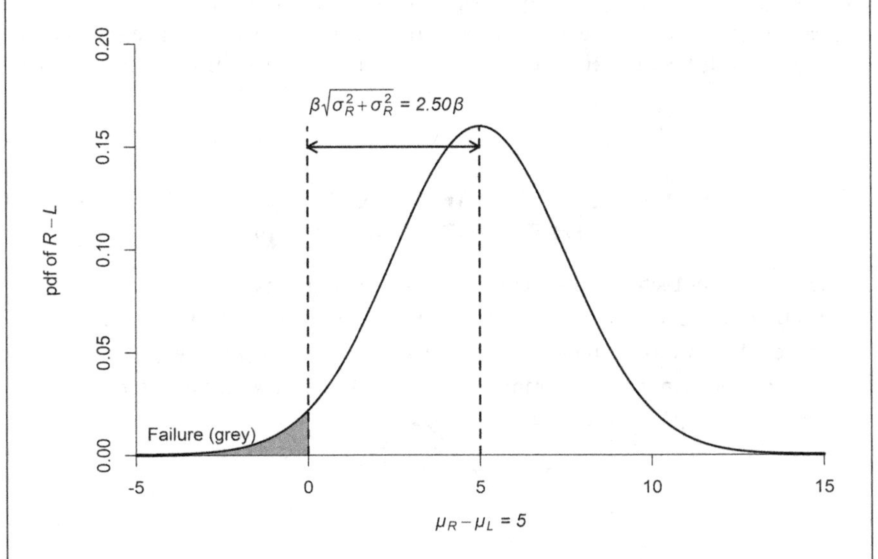

Probability of failure (grey area under the left part of the curve) for vertical pile example.

7.3 RELIABILITY INDEX OF NON-LINEAR SYSTEMS

7.3.1 First Order Second Moment (FOSM) Method

For non-linear systems the reliability index can be approximated by using first- and second-order Taylor approximations directly to find the mean and variance of the non-linear limit state function as discussed in section 5.3.1. Consider a non-linear limit state function $g(R,L)$ where resistance R and load L are both normally distributed and independent random variables with mean μ_R, μ_L and variance σ_R^2, σ_L^2. A direct first-order Taylor approximation of the limit state function evaluated at the mean value of the R and L gives:

$$\mu_g = E(g(R,L)) \approx g(\mu_R, \mu_L) \tag{7.13}$$

$$\sigma_g^2 = V(g(R,L)) \approx \left(\frac{\partial g(\mu_R,\mu_L)}{\partial R}\right)^2 \sigma_R^2 + \left(\frac{\partial g(\mu_R,\mu_L)}{\partial L}\right)^2 \sigma_L^2$$

In this case the reliability index is simply defined by combining the definition in Eq. 7.5 with the expressions in Eq. 7.13. When using Taylor approximations of first order, this method is often referred to in the technical literature as the First Order Second Moment (FOSM) method, where the second-order moment refers to the use of the mean and the variance. If second-order terms are included in the Taylor approximations, for example to get a more precise approximation of the mean value, the method is referred to as Second Order Second Order Moment (SOSM) method. While intuitively appealing and easy to use, this method has been criticised for use in reliability engineering because it relies on the exact form of the limit state function.

EXAMPLE 7.2 INVARIANCE TO THE FORM OF THE LIMIT STATE FUNCTION

Consider a geotechnical problem involving a vertical cut of height $H = 10m$ in a purely cohesive soil as shown in the figure below with a sliding failure surface inclined 45°. In this example, the soil cohesion c and the unit weight of soil γ are considered independent random variables. All other variables are fixed.

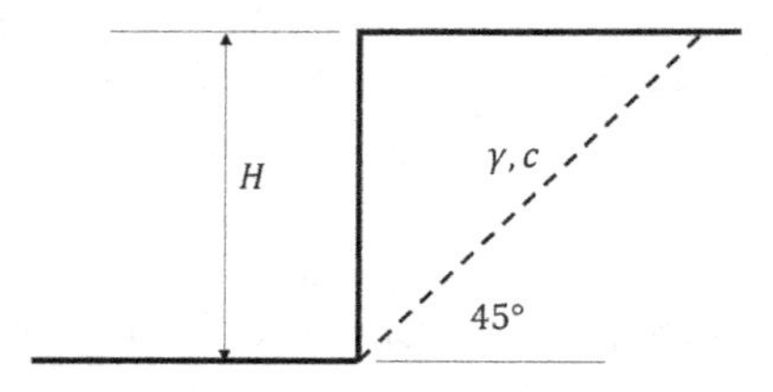

Vertical cut in slope.

Considering the balance between the shear stresses along the failure surface gives a limit state function as:

$$g(c,\gamma)=c-\frac{H\gamma}{4}$$

where cohesion c and unit weight γ are both considered independent random variables. The cohesive strength of the soil has a mean value of $\mu_c = 100KPa$ and a standard deviation of $\sigma_c = 30KPa$. The unit weight of the soil has a mean value of $\gamma = 20KN/m^3$ and a standard deviation of $\sigma_\gamma = 2KN/m^3$. To calculate the reliability index using the FOSM method requires the mean value and variance of the limit state function to be evaluated using first-order Taylor approximations evaluated in the mean values of c and γ as:

$$\mu_g = E(g(c,\gamma)) \approx \mu_c - \frac{H\mu_\gamma}{4} = 100KPa - \frac{10m \times 20KN/m^3}{4} = 50KPa$$

$$\sigma_g^2 = V(g(c,\gamma)) \approx 1^2\sigma_c^2 + \left(\frac{H}{4}\right)^2 \sigma_\gamma^2 = (30KPa)^2 + \left(\frac{10m}{4}\right)^2 (2KN/m^3)^2 = 925KPa^2$$

as: $\frac{\partial g}{\partial c} = 1$ and $\frac{\partial g}{\partial \gamma} = \frac{H}{4}$

Finally, the reliability index is defined according to Eq. 7.5 as:

$$\beta = \frac{\mu_g}{\sigma_g} = \frac{50KPa}{\sqrt{925KPa^2}} = 1.644$$

Next, consider the same problem but now let the limit state function reflect the ratio between the forces, i.e.:

$$g(c,\gamma) = \frac{4c}{H\gamma} - 1$$

Again, using first-order Taylor approximations evaluated in the mean values of c and γ now gives:

$$\mu_g = E(g(c,\gamma)) \approx \frac{4\mu_c}{H\mu_\gamma} - 1 = \frac{4 \times 100KPa}{10m \times 20KN/m^3} - 1 = 1.0$$

$$\sigma_g^2 = V(g(c,\gamma)) \approx \left(\frac{4}{H\mu_\gamma}\right)^2 \sigma_c^2 + \left(-\frac{4\mu_c}{H\mu_\gamma^2}\right)^2 \sigma_\gamma^2$$

$$= \left(\frac{4}{10m \times 20KN/m^3}\right)^2 (30KPa)^2 + \left(-\frac{4 \times 100KPa}{10m \times (20KN/m^3)^2}\right)^2 (2KN/m^3)^2$$

$$= 0.40$$

Finally, the reliability index is once again calculated according to Eq. 7.5 as:

$$\beta = \frac{\mu_g}{\sigma_g} = \frac{1.0}{\sqrt{0.40}} = 1.581$$

Notice that even for this very simple example, the resulting reliability index is different for two seemingly completely identical evaluations of the same problem. Thus, the reliability of the system depends on the problem formulation, which is not a desirable property.

7.3.2 The Hasofer-Lind Reliability Index Method (FORM)

A possible solution to the invariance problem described above was proposed by Hasofer and Lind in 1974 (hence the Hasofer-Lind method), also sometimes known as the First Order Reliability method (FORM), which is not to be confused with the FOSM method introduced in the previous section. FORM starts by proposing the existence of a *design point* on the limit state boundary between a failure and a safe system state. As the location of this design point is not known in advance, an iterative procedure is required to find the reliability index. First, consider a limit state function evaluated at the design point:

$$g\left(x_1^*, x_2^*, \cdots, x_n^*\right) = 0 \tag{7.14}$$

Here $x_i^*, i = 1, \ldots, n$ represents the (as yet unknown) design point and is a series of realisations of random variables $x_i^*, i = 1, \ldots, n$. For ease of presentation, the random variables are considered independent and with mean values $(\mu_1, \mu_2, \ldots, \mu_n)$and variance $\left(\sigma_1^2, \sigma_2^2, \ldots, \sigma_n^2\right)$, respectively.

Next, consider the standardised form, Z_i, of each of the n input variables as

$$Z_i = \frac{X_i - \mu_i}{\sigma_i}, i = 1, 2, \cdots, n \tag{7.15}$$

Geometrically, the reliability index can still be defined as the shortest geometrical distance between the origin of the normalised variables to the design point on the limit state function border between failure and safety defined as $g(z_1, z_2, \ldots, z_n)$ as shown in Figure 7.2. This geometric distance is defined mathematically as:

$$\beta = \min\left(\sqrt{\sum_{i=1}^{n} z_i^2}\right) \tag{7.16}$$

subject to the solution satisfying that the limit state function equals zero, i.e. $g(z_1, \ldots, z_n) = 0$.

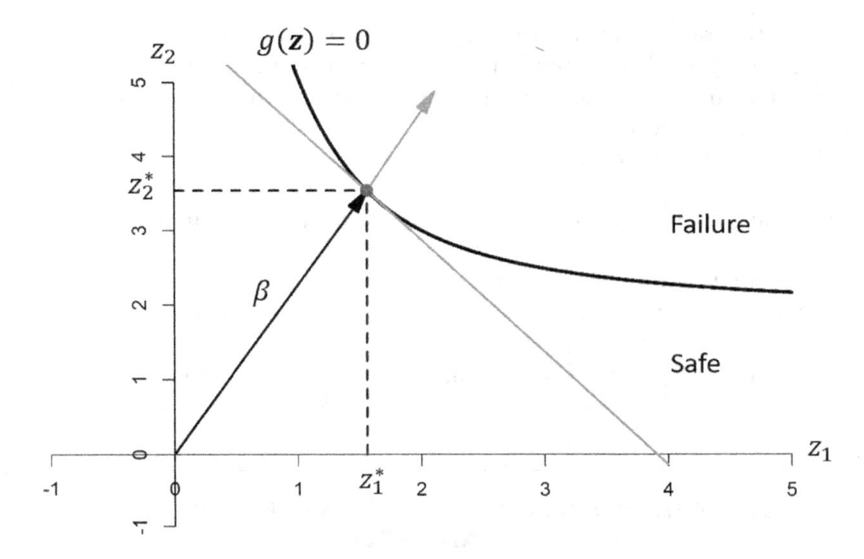

Figure 7.2 Definition of the reliability index based on the Hasofer-Lind method.

Calculating the reliability index is now reduced to a constrained optimisation problem, finding the minimum distance conditional on the solution being on the limit state border between failure and safety. Solving constrained optimisation problems can be complex, but an iterative procedure is readily available in the form of the Hasofer-Lind method.

The outward directed normal vector to the failure surface at the design point $\left(z_1^*, z_2^*\right)$ is defined as:

$$\alpha_i = \left(\frac{\partial g}{\partial z_1}, \cdots, \frac{\partial g}{\partial z_n}\right) \tag{7.17}$$

From a geometric consideration of Figure 7.2 it can be observed that the two design values z_1^* and z_2^* are directional cosine values for a vector with length β (the reliability index) and an angle defined by the normal vector $\boldsymbol{\alpha}$ and therefore:

$$z_i^* = \beta \alpha_i, i = 1, \cdots, n \tag{7.18}$$

where α_i is the ith element in the normalised vector, i.e.:

$$\alpha_i = \frac{-\dfrac{\partial g\left(z_i^*\right)}{\partial z_i}}{\sqrt{\displaystyle\sum_{j=1}^{n}\left(\frac{\partial g\left(z_j^*\right)}{\partial z_j}\right)^2}}, \; i = 1, \cdots, n \tag{7.19}$$

The iterative procedure required to find the reliability index consists of the five steps outlined below. But before starting the iterations it is necessary to rewrite the limit state function in terms of the normalised variables, such that $g(x_i,\ldots,x_n)$ becomes $g(z_i,\ldots,z_n)$.

The five steps of the iteration are as follows:

1. Assume initial values of the design point x_i^* and compute the corresponding values of the normalised variables z_i^*.
2. Compute the values of α_i at z_i^* using Eq. 7.19.
3. Substitute values of $z_i = \alpha_i\beta$ into the limit state function $g(z_i,\ldots,z_n)$ and solve for β.
4. Use the new value of β to update values of design point as $z_i^* = \alpha_i\beta$.
5. Repeat steps 2 to 5 until β converges sufficiently.

EXAMPLE 7.3 RELIABILITY OF A VERTICAL CUT EMBANKMENT

Consider again the limit state function for a vertical cut in cohesive soil expressed as a factor of safety:

$$g(c,\gamma) = \frac{4c}{H\gamma} - 1$$

First, express the limit state function according to the normalised parameters z_c and z_γ:

$$g(c,\gamma) = \frac{4(\mu_c + z_c\sigma_c)}{H(\mu_\gamma + z_\gamma\sigma_\gamma)} - 1$$

Step 1: Assume a set of initial values for the two random variables c and γ. Assume that the initial values equal the mean values and compute the corresponding normalised values, i.e.:

$$c^* = \mu_c \rightarrow z_c^* = \frac{\mu_c - \mu_c}{\sigma_c} = 0$$

$$\gamma^* = \mu_\gamma \rightarrow z_\gamma^* = \frac{\mu_\gamma - \mu_\gamma}{\sigma_\gamma} = 0$$

Step 2: Compute values of α_c and α_γ at the values of z_c^* and z_c from step 1:

$$\frac{\partial g}{\partial z_c} = \frac{4\sigma_c}{H(\mu_\gamma + z_\gamma\sigma_\gamma)}$$

$$\frac{\partial g}{\partial z_\gamma} = \frac{-4(\mu_c + z_c\sigma_c)}{H(\mu_\gamma + z_\gamma\sigma_\gamma)^2}\sigma_\gamma$$

Using these differentials gives the following values of α:

$$\alpha_c = \frac{-\dfrac{4\times 30KPa}{10m\times\left(20KN/m^3+0\times 2KN/m^3\right)}}{\sqrt{\left(\dfrac{4\times 30KPa}{10m\times\left(20KN/m^3+0\times 2KN/m^3\right)}\right)^2 + \left(\dfrac{-4\times(100KPa+0\times 30Kpa)\times 2KN/m^3}{10m\times\left(20KN/m^3+0\times 2KN/m^3\right)^2}\right)^2}} = -0.9487$$

$$\alpha_\gamma = \frac{-\dfrac{4\times(100KPa+0\times 30KPa)\times 2KN/m^3}{10m\times\left(20KN/m^3+0\times 2KN/m^3\right)^2}}{\sqrt{\left(\dfrac{4\times 30KPa}{10m\times\left(20KN/m^3+0\times 2KN/m^3\right)}\right)^2 + \left(\dfrac{-4\times(100KPa+0\times 30Kpa)\times 2KN/m^3}{10m\times\left(20KN/m^3+0\times 2KN/m^3\right)^2}\right)^2}} = 0.3162$$

Step 3: Substitute values of $z = \alpha\times\beta$ into the limit state function and isolate β:

$$g\left(z_c, z_\gamma\right) = \frac{4\left(\mu_c + \alpha_c\beta\sigma_c\right)}{H\left(\mu_\gamma + \alpha_\gamma\beta\sigma_\gamma\right)} - 1$$

From which β can be isolated as:

$$\beta = \frac{H\mu_\gamma - 4\mu_c}{4\alpha_c\sigma_c - H\alpha_\gamma\sigma_\gamma}$$

Using the values of α_c and α_γ from Step 2 gives:

$$\beta = \frac{10m\times 20KN/m^3 - 4\times 100KPa}{4\times(-0.9487)\times 30KPa - 10m\times 0.3162\times 2KN/m^3} = 1.6644$$

Step 4: Use new value of β to update the design point:

$$z_c^* = \alpha_c\beta = -0.9487\times 1.6644 = -1.5790$$

$$z_\gamma^* = \alpha_\gamma\beta = 0.3162\times 1.6644 = 0.5263$$

This completes the first iteration. The updated design values from Step 4 are now used as input into Step 2. The iterative procedure is repeated until the reliability index β is judged to have converged sufficiently. A summary of the iterations is shown in the table below. As can be observed, the procedure has converged after only three iterations.

Expression	1. Iteration	2. Iteration	3. Iteration
	Step 2:		
$\frac{\partial g}{\partial Z_c}$	0.60	0.57	0.5842
$\frac{\partial g}{\partial Z_g}$	−0.20	−0.095	−0.0974
α_c	−0.9487	−0.9864	0.9864
α_g	0.3162	0.1640	0.1644
	Step 3:		
β	1.6644	1.6440	0.1644
	Step 4:		
Z_c	−1.5790	−1.6216	−1.6216
Z_g	0.5263	0.2703	0.2703

Results from iteration using the Hasofer-Lind method.

Alternatively, the reliability index β could have been determined using EXCEL's (or any other computational software's) in-built ability to solve constraint optimisation problems.

7.4 SOLVE FORM USING EXCEL'S CONSTRAINED OPTIMISATION SOLVER

This section introduces a more direct method of estimating the reliability index using the built-in optimisation tool available in EXCEL known as the Solver. Before starting, it might be necessary to activate the Solver, which can be done by following the guidance in Figure 7.3. Once the Solver has been initiated, the reliability index can be found by solving Eq. 7.5 directly following the steps below; see the associated steps in Figure 7.4.

7.4.1 Introducing the EXCEL Solver

To introduce the Solver and its basic functionalities, consider first a simple mathematical optimisation problem of finding the value of a variable x that maximises the function:

$$f(x) = -x^2 + x + 3$$

Mathematically, this optimisation problem can be written as:

$$\max_{x \in \mathcal{R}} \{f(x)\}$$

which means searching for the maximum value of the function $f(x)$ among all possible values of the input variable x. The function $f(x)$ is now defined as the *objective function*. This is an unconstrained optimisation problem as the space of possible numbers includes all real numbers, i.e. there are no constraints on the input variables. To solve this optimisation problem in EXCEL, follow the steps below.

Step 1: Set up the optimisation problem in EXCEL. This is the most difficult part and requires understanding of the problem at hand. Define a cell for the input (cell B2 on Figure 7.4) and a cell for the associated function value (cell B3). An initial start guess of the optimal value should be provided in cell B2, but this is only a first guess and does not have to be the correct solution. Start the Solver and:

- Select the cell containing values of the objective function; cell B3 in this case as this cell contains values of Eq. 7.5 (our objective is to minimise Eq. 7.5).
- Select what optimisation operation Solver should attempt; the options include maximising, minimising, or finding an exact value of the objective function (cell B3).
- Finally, select the input value over which the search should be conducted (cell B2).

Step 2: Once the Solver problem has been completed, press the *Solve* button. At this point, the internal optimisation algorithm in EXCEL will attempt to find the value of the input parameter in cell B2 that solves the defined optimisation problem. Once completed, EXCEL will report whether it was able to find a solution to the problem and give the user the option of keeping the solution or default to the initial values.

Step 3: If the solution is deemed acceptable, press *OK* and the optimal value will be automatically transferred to cell B2 and the corresponding maximum value of the function updated accordingly in cell B3. At this point, the optimisation is complete. As can be seen from Figure 7.4, the Solver has found that a value of $x = 0.50$ gives the maximum value of the objective function.

7.4.2 Constrained Optimisation

Optimisation problems involving constraints on the possible solutions can be much more complicated than unconstrained problems, and this is where the Solver can be a very powerful tool for the analyst. To add constrains,

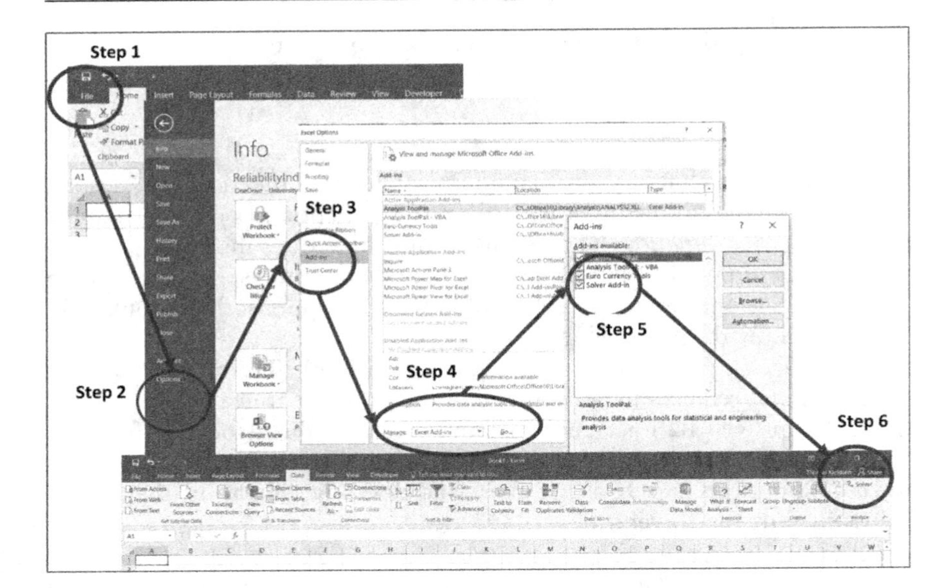

Figure 7.3 Screenshots showing the steps necessary to activate the EXCEL solver.

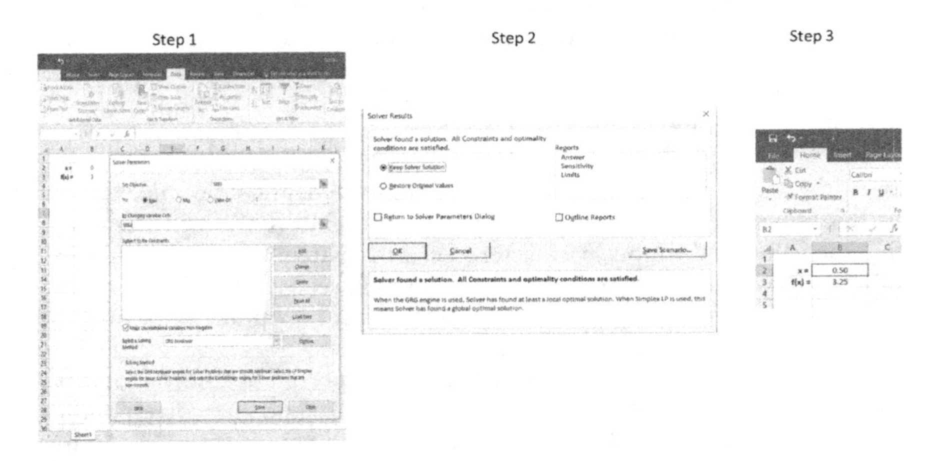

Figure 7.4 Three steps required for solving a simple unconstrained optimisation problem.

simply press the *Add* button in the constraints section of the Solver interface as shown in Figure 7.5.

Once the constraints have been defined, the Solver will only search for solutions that comply with the imposed constraints.

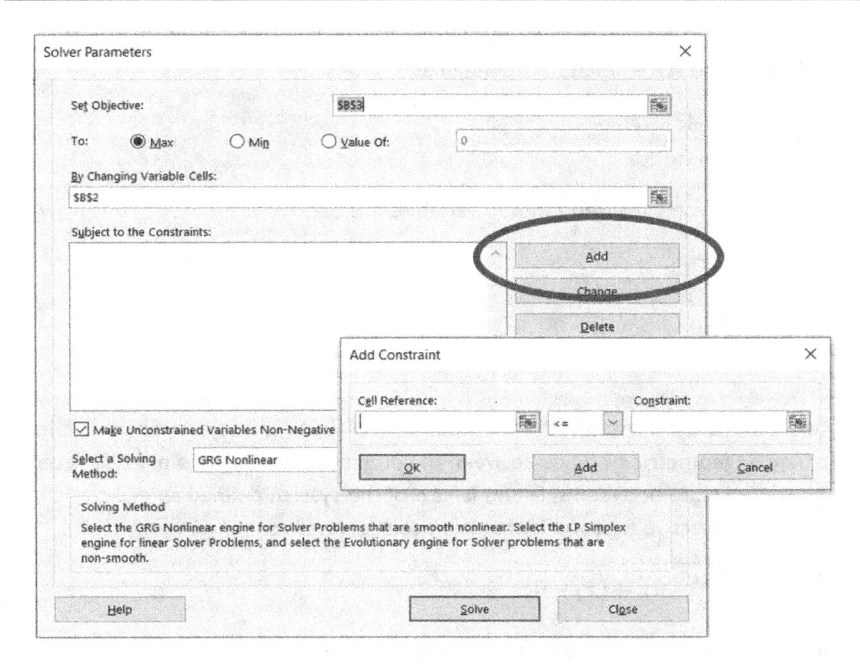

Figure 7.5 Add constraints to a Solver solution.

EXAMPLE 7.4 RELIABILITY OF A VERTICAL CUT EMBANKMENT: EXCEL SOLUTION

To illustrate the use of multivariate constrained optimisation, consider the problem of calculating the reliability index of a vertical cut into a purely cohesive soil introduced in Examples 7.2 and 7.3. As before, the unit weight of the soil material γ and the cohesion c are both considered random variables with mean and standard deviation given in Example 7.2 and repeated in the table below. The height of the cut is $H = 10m$ and the failure surface is assumed to have a 45-degree incline.

Mean and standard deviation of unit weigh and cohesion

Variable	*Symbol*	*Unit*	*Mean*	*Std Dev*
Unit weight	γ	kN/m^3	100	30
Cohesion	c	kPa	20	2

The limited state function is defined as:

$$g(c,\gamma) = \frac{4c}{H\gamma} - 1$$

Next, the two normalised random variables are defined as:

$$z_c = \frac{c - \mu_c}{\sigma_c}$$

$$z_\gamma = \frac{\gamma - \mu_c}{\sigma_\gamma}$$

Referring to the Hasofer-Lind method, the reliability index is defined as the minimum geometric distance between the origin to the design point (z_c^*, z_γ^*) on the limited state between safe and failure of the system defined as $g(c,\gamma) = 0$. Thus, the objective function is the distance d defined as:

$$d = \sqrt{(z_c^* - 0)^2 + (z_\gamma^* - 0)^2}$$

But the solution is constrained to values of z_c and z_γ that are located on the space defined by $g(c,\gamma) = 0$.

The use of Solver to find the minimum distance is illustrated in the figure below and follows the three steps outlined above.

Step 1: Set up the optimisation problem in an EXCEL sheet.

- Define the input parameters: cells B8 and B9 and the associated normalised values in cells F8 and F9.
- Define the objective function (cell B16), the distance between the origin to the design point.
- Define the constraint as the value of the limited state function: cell B12.

Step 2: Activate the Solver function and define.

- Objective: select cell B16 and ensure to select the minimisation goal.
- Set the "Changing Variables Cells" to the input values in cells B8 and B9.
- Add a constraint to ensure that limited state function in cell B12 equals zero.
- Press Solve and accept the solution.

Step 3: The values of c and γ resulting in the minimum distance are now reported and the resulting reliability index is found as $\beta = 1.64$. Compare this result to the solution of the same problem in Example 7.3.

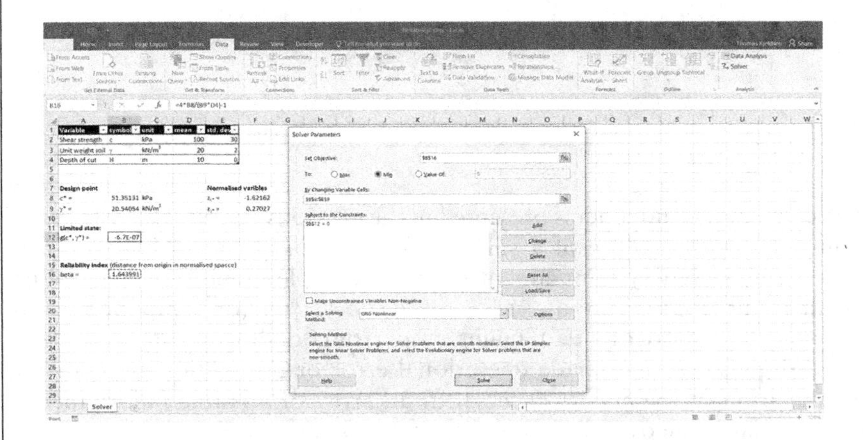

Solution of Eq. 7.5 to estimate the reliability index directly using the multivariate constrained optimisation functionality of the EXCEL Solver.

Chapter 8

Complex Systems

The discussions and models presented in the previous chapters have focussed primarily on the failure of a single system or component. However, real-world engineered systems are often more complex, and a system-failure might result from failure of one or more subsystems (or components). For example, the failure of a dam might result from flooding or internal erosion or earthquake or man-made operational error or foundation error, or any combination of the above. Faced with an array of possible failure mechanisms, what is the overall probability of a system failure?

To answer such questions, it is necessary to study the failure of system. Three fundamental types of systems can be defined: a series system, a parallel system, and a binomial system. These systems can then be joined together to represent more complex systems.

8.1 SERIES SYSTEMS

A series system consists of n subsystems connected in such a way that if one subsystem fails, then the entire system fails. Therefore, the entire system is only as strong as the weakest subsystem. For example, a chain will break (system failure) as soon as the weakest link breaks. A schematic of a series system consisting of n components is shown in Figure 8.1.

To calculate the probability of failure P_f for the entire system, an indicator function is defined for each component $i = 1,\ldots,n$ such that:

$$F_i = \begin{cases} 0 & \text{component in a no-failure state} \\ 1 & \text{component in a failure state} \end{cases} \tag{8.1}$$

From Eq. 8.1 the probability of a failure in the i th component is $p_i = P\{F_i = 1\}$ and the probability of no failure is $P\{F_i = 0\} = 1 - P\{F_i = 1\} = (1 - p_i)$.

DOI: 10.1201/9781032700373-8

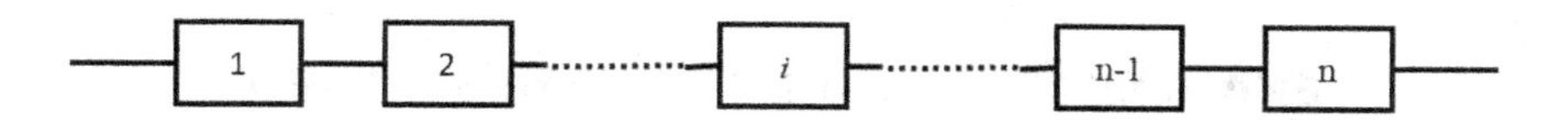

Figure 8.1 Schematic of series system consisting of *n* components.

Recognising that the entire system fails if a single component fails is the same as system survival only if no component fails, i.e.:

$$\begin{aligned} P_f &= (\text{System failure}) = 1 - P(\text{System survival}) \\ &= 1 - P((F_1 = 0) \cap (F_2 = 0) \cap \ldots \cap (F_n = 0)) \end{aligned} \tag{8.2}$$

The system survival part of the equation (non-failure) can be interpreted as the probability of: (survival of component 1) AND (survival of component 2) AND ... AND (survival of component n). If the components fail independently of each other, then Eq. 8.2 becomes:

$$P_f = 1 - \prod_{i=1}^{n} (1 - p_i) \tag{8.3}$$

In the special case where the probability of failure is the same for all components $p_i = p$, then Eq. 8.3 reduces further to:

$$P_f = 1 - (1 - P)^n \tag{8.4}$$

Thus, for the special case where all components fail or survive independently, and all have the same probability, the overall probability of failure of the entire system depends only on the number of components n and the individual failure probability p. The relationship between P_f, n, and p is shown graphically in Figure 8.2.

The figure shows that as the number of components increases, so does the probability of failure of the entire system, and that for very high individual failure probabilities, the probability of overall system failure converges towards one regardless of the number of components. In other words, adding additional elements will increase the overall probability of system failure.

8.2 PARALLEL SYSTEMS

A parallel system is designed such that a system failure will only occur if all sub-subsystems (components) fail simultaneously, and therefore such systems are often preferable to series systems if possible. For example, you need to buy milk from one of five possible shops. As long as just one shop is open, you can buy the milk, and only if all shops are closed can you not get any milk (a system failure). A schematic of a parallel system is shown in Figure 8.3.

EXAMPLE 8.1 FAILURE OF A SERIES SYSTEM

Two cities are connected by a railway line crossing over three rivers as shown below.

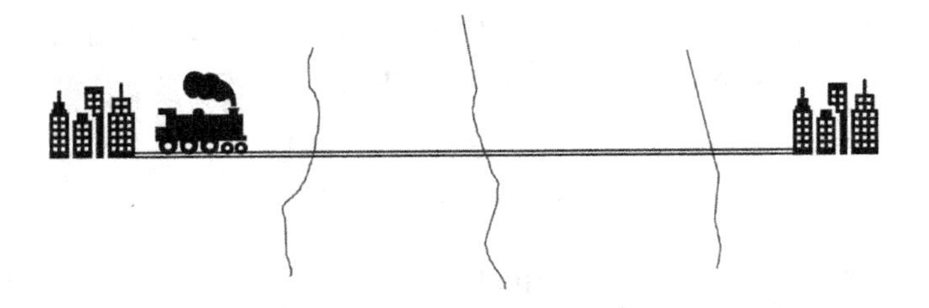

At each crossing a culvert has been constructed with a capacity reducing the probability of flooding at each site to 0.10 each year. If flooding occurs at any point on the line, then the train is unable to get from one city to the next resulting in a system failure. Assuming the rivers flood independently of each other, what is the probability of a system failure?

Using Eq. 8.4 the probability of a system failure P_f for a system consisting of $n = 3$ independent components each with a failure probability at each river crossing of $p = 0.10$ is given as:

$$P_f = 1 - (1 - 0.1)^3 = 0.2710$$

Maintaining the indicator function from Eq. 8.1 the probability of failure P_f of the parallel system can be interpreted as the probability of: (failure of component 1) AND (failure of component 2) AND ... AND (failure of component n), i.e.:

$$P_f = P\left((F_1 = 1) \cap (F_2 = 1) \cap \ldots \cap (F_n = 1)\right) \tag{8.5}$$

If components fail independently of each other, then this expression reduces to:

$$P_f = \prod_{i=1}^{n} p_i \tag{8.6}$$

and if the failure probability is the same for all components $p_i = p$ then this further reduces to:

$$P_f = p^n \tag{8.7}$$

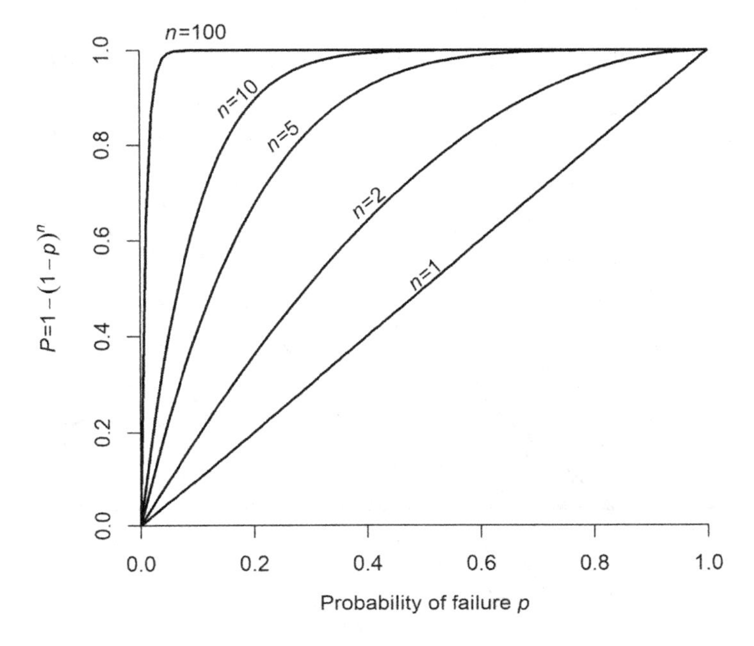

Figure 8.2 Probability of system failure plotted against failure probability of individual components.

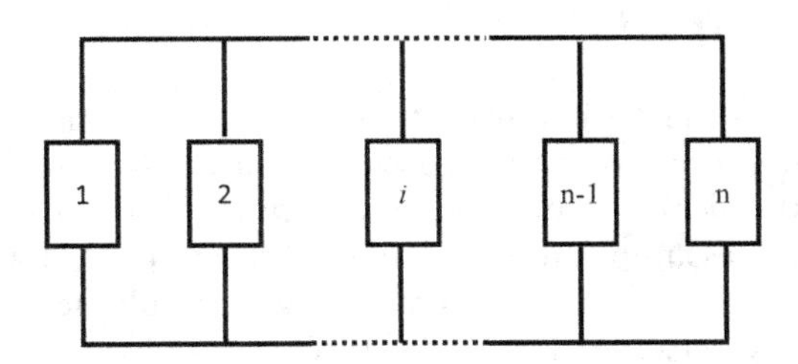

Figure 8.3 Schematic of parallel system consisting of n components.

Thus, for the special case where components fail independently and have the same failure probability, the probability of failure of the entire system depends only on the number of components and the probability of failure of each component as illustrated in Eq. 8.7 and Figure 8.4.

In contrast to a series system, a parallel system becomes more robust (lower probability of system failure) the more components are added to the system. But again, if the individual failure probability is sufficiently high, then the system failure approaches 1 for the entire system.

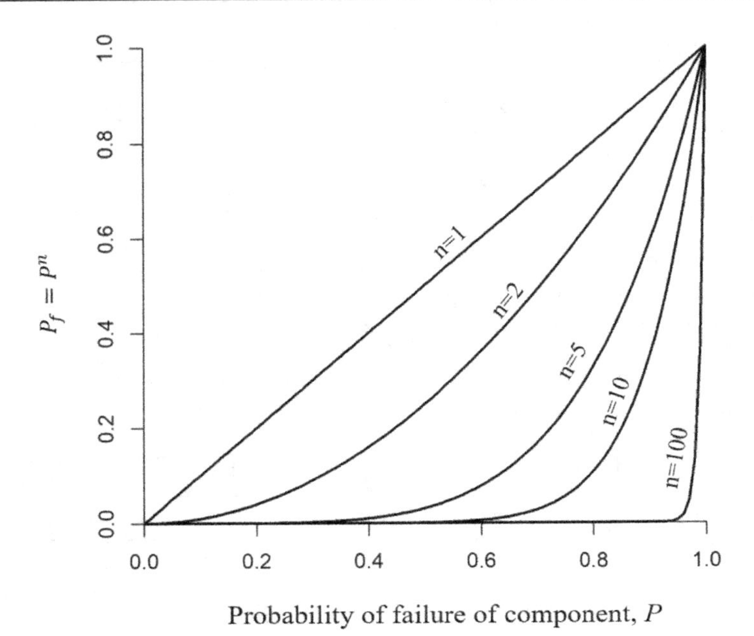

Figure 8.4 Probability of system failure p_f plotted against failure probability of individual components.

8.3 PROBABILITY BOUNDS FOR SERIES AND PARALLEL SYSTEMS

The derivations of the probability of failure for series and parallel systems presented in the preceding sections assume that the individual components (or subsystems) fail or survive independently of each other. However, for such systems this assumption is violated. For example, a large earthquake might be expected to destroy several bridges simultaneously. If positive correlation exists between the reliability of the individual elements, then it is possible to provide a set of bounds for the probability of failure for the entire system representing the cases of perfectly correlated or perfectly uncorrelated (independent) performance of elements. For a series system with perfectly correlated performance of the elements (all elements fail as soon as one element fails), the probability of failure is equal to the largest probability of failure of the individual elements. Consider a series system consisting of n elements each assigned a probability of failure P_i, $i = 1, \ldots n$. In this case, the bounds on the probability of failure of the entire system, P_f, is given as:

$$\max_{1 \le i \le n}(p_i) \le P_f \le 1 - \prod_{i=1}^{n}(1 - p_i) \qquad (8.8)$$

where the upper bound is the case where all elements are independent of each other. Similarly, for a parallel system the probability bounds on the system probability of failure is given as:

$$\prod_{i=1}^{n} p_i \le P_f \le \min_{1\le i\le n}(p_i) \tag{8.9}$$

Here the lower bound represents the case of independent elements while the upper bound represents the case where the element with the smallest individual failure probability determines the upper bound; remember a parallel system only fails once all elements have failed.

8.4 BINOMIAL SYSTEMS

Binomial systems are also known as m-out-of-n systems, meaning that they fail if m out of the total of n components fail, regardless of which of the m systems fail. For example, a car fails to complete a journey if any one or more of its four wheels puncture.

Assume as above that the components fail independently and that the failure probability p is identical for each component. Defining a random variable M as the number of components that fail, then the probability of $M = m$ failures can be represented by the probability mass function (pmf) of a binomial distribution (see section 2.6.7):

$$f(m) = P(M = m) = \binom{n}{m} P^m (1-P)^{n-m} \tag{8.10}$$

where $\binom{n}{m} = \frac{n!}{m!(n-m)!}$ is the binomial coefficient and represent the different possible combinations to select m elements out of the n total. The total system failure occurs if m or more components fail, i.e. $M \ge m$. Therefore, the probability of a system failure is:

$$\begin{aligned} P_f &= P(M \ge m) \\ &= P(M = m) + P(M = (m+1)) + \cdots + P(M = n) \\ &= 1 - P(M = (m-1)) - \cdots - P(M = 0) \\ &= 1 - \sum_{i=0}^{m-1} f(i) \end{aligned} \tag{8.11}$$

where f is the pmf of the binomial distribution defined in Eq. 8.10 above. The special case of the binomial system where $m = n$ represents a parallel system.

EXAMPLE 8.2 BINOMIAL SYSTEM

A contractor has a workforce of $n = 20$ people. At least 16 people need to be available for work on a particular day for work to be undertaken. Assume that the probability for a person to turn up for work on any given day outside holiday seasons is the same across all individuals people and set to 0.85, and that individuals do not influence each other's decision to work or not. What is the probability that work cannot progress on any given day?

First, the random variable M is defined as the number of workers not turning up for work on any given day. Next, a system failure is defined as a day where work cannot be undertaken, which occurs if five or more individuals do not turn up for work, $M \geq 5$. Therefore, using Eq. 8.11 with $n = 20$ and the probability of a failure of an individual to turn up for work set to $p = 1 - 0.85 = 0.15$, the probability of a system failure P_f is defined as:

$$P_f = P(M \geq 5) = 1 - \sum_{i=0}^{5-1} f(i)$$

where $f(i)$ is the binomial distribution evaluated for i

$$f(i) = \binom{n}{i} p^i (1-p)^{n-i}$$

Evaluation of this equation requires the pmf of the binomial distribution to be evaluated for $i = 0, 1, 2, 3, 4$ and then summed-up. The pmf is shown in the figure below, where the outcomes that result in a system failure ($i \geq 5$) are highlighted in grey and the outcomes that will allow work to continue ($i < 5$) are highlighted in black.

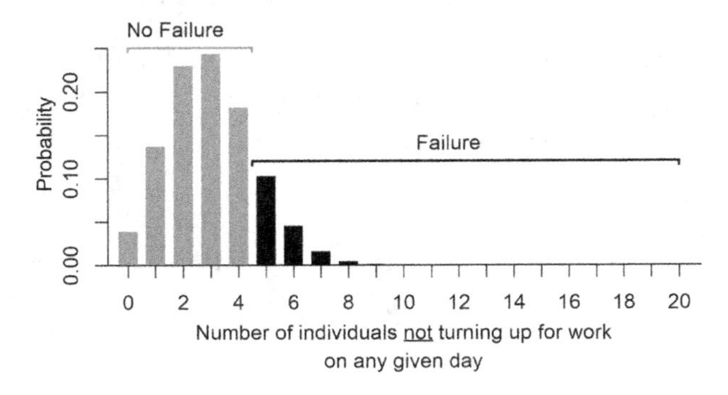

Probability mass function (pmf) of binomial distribution showing the probability of i individuals not turning up for work.

Finally, the calculations needed to establish the probability of a system failure are shown in the table below.

Evaluation of the pmf and the cdf for the Binomial Distribution

i	$\binom{n}{i}$	$f(i)$	$\Sigma f(i)$
0	1	0.0388	0.0388
1	20	0.1368	0.1756
2	190	0.2293	0.4049
3	1140	0.2428	0.6477
4	4845	0.1821	0.8298

The last column in the table is the cumulative sum of $f(i)$, and the probability of failure is therefore calculated as:

$$P_f = 1 - \sum_{i=0}^{5-1} f(i) = 1 - 0.8298 = 0.1702$$

8.5 MORE COMPLEX SYSTEM CONFIGURATIONS

The systems presented in the previous sections can be considered the basic building blocks of many more complex real-world systems. For critical infrastructure it is important that the entire system does not fail based on the performance of a single component. Instead, systems should ideally be designed either as a series of parallel systems where the individual components within the parallel subsystems can act as safety features preventing complete system failure resulting from failure of a single critical component.

The system shown in Figure 8.5 can be recognised as a series of m subsystems forming a series-system. Each subsystem consists of $n_i, i = 1,\ldots,m$ components forming a parallel system. The probability of failure of each individual component is defined as $p_{ij}, i = 1,\ldots,m$ and $j = 1,\ldots,n_i$. Assuming components fail independently, determining the probability of a complete system failure P_f starts from Eq. 8.3 which characterise the series system consisting of m subsystems, each with its own probability of failure $P_{f,i}$:

$$\begin{aligned} P_f &= 1 - \prod_{i=1}^{m} (\text{Probability of no failure of sub-system } i) \\ &= 1 - \prod_{i=1}^{m} \left(1 - P_{f,i}\right) \end{aligned} \tag{8.12}$$

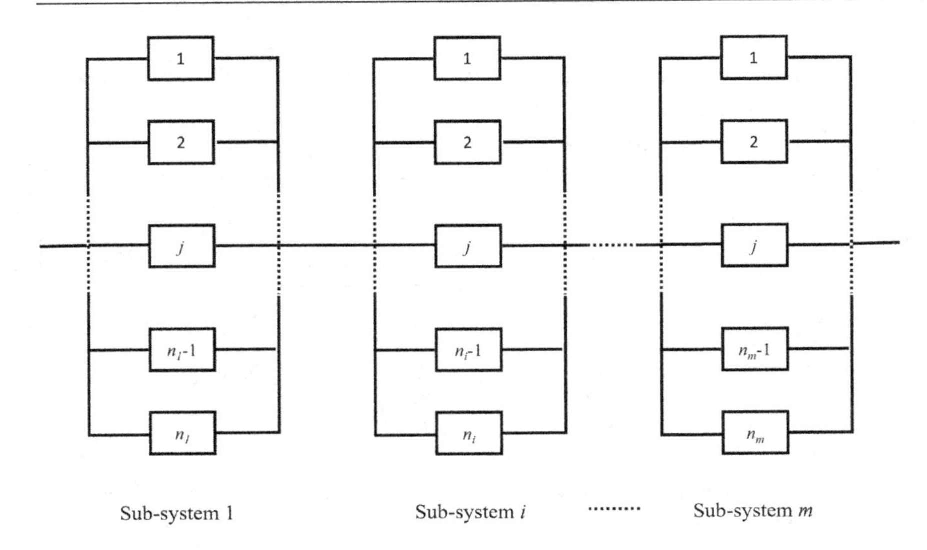

Figure 8.5 Schematic of a series system consisting of $i = 1,\ldots,m$ parallel systems, each consisting of n_i components.

The probability of failure of the i th subsystem with n_i parallel elements with failure probability p_{ij} is determined from Eq. 8.6 as:

$$P_{f,i} = \prod_{i=1}^{n_i} p_{ij} \tag{8.13}$$

Combining Eqs. 8.12 and 8.13 gives the final expression for the probability of system failure for the series-parallel system in Figure 8.5 as:

$$P_f = 1 - \prod_{i=1}^{m} \left(1 - \prod_{j=1}^{n_i} p_{ij} \right) \tag{8.14}$$

For complex real-world systems, these calculations can quickly become cumbersome, and the next section will introduce fault trees as a way of managing risk analysis of complex systems.

EXAMPLE 8.3 FAILURE OF MORE COMPLEX SYSTEMS

Two locations are connected via roads crossing three bridges (1,2 and 3) as shown in the figure below.The area is characterised by high seismic activity and the probability of each bridge failing in case of a large earthquake is denoted p_1, p_2 and p_3. Find the probability of not being able to travel between the two locations following a large earthquake.

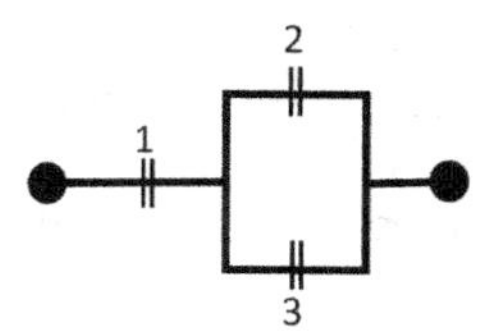

Solution: The complex system is made up of $m = 2$ parallel subsystems connected as a series system. The first subsystem consists of only one bridge (1), while the second parallel system consists of two bridges. The probability of failure of each subsystem is calculated using Eq. 8.6:

Subsystem 1: $P_{f,1} = p_1$

Subsystem 2: $P_{f,2} = p_2 \times p_3$

Substituting $P_{f,1}$ and $P_{f,2}$ into the general formula for the probability of failure for a complex system in Eq. 8.12 gives:

$$\begin{aligned} P_f &= 1 - (1 - P_{f,1})(1 - P_{f,2}) \\ &= 1 - (1 - p_1)(1 - p_2 p_3) \\ &= p_1 + p_2 p_3 - p_1 p_2 p_3 \end{aligned}$$

8.6 FAULT TREE ANALYSIS

Fault tree analysis is a graphical technique that can help build understanding of how a combination of individual causes, or events (e.g. equipment failure, human errors), can ultimately cause an undesirable critical system event (a failure). The method follows a top-down approach developing a visual representation of the possible pathways between basic events and a particular undesirable failure, denoted the top event. The basic events ultimately causing the top event are connected using logic symbols (AND, OR, etc.), referred to as gates with reference to electronic systems. Conceptually, the AND and OR gates can also be considered as parallel and series systems, respectively.

To develop a fault tree in its most basic form, it is necessary to first develop a thorough understanding of the system and then go through the following steps:

- Define the undesirable event (top event) of the system. This event will be at the top of the fault tree.
- Identify all the basic events that could occur and contribute to the undesired event.

- Construct the fault tree by starting with the top event and adding each basic event and its causes as branches. Use logical operators such as AND and OR to describe the relationships between the events.
- Identify minimum cut sets.
- Assign failure probabilities to elements of events.

8.6.1 Fault Tree Symbols and Definitions

Visually, a fault tree consists of a number of events connected via logical gates. In the following, the most basic components of a fault tree are discussed, including their graphical representation (see summary in Figure 8.6) and the basic probabilistic relationships between events and logical gates.

8.6.1.1 Top Event

The top event is located at the top of the fault tree and represents the subject of the analysis, e.g. a dam failure. This event is shown visually by a rectangle with an input but no output as it represents the culmination of events in the entire fault tree.

8.6.1.2 Intermediate Event

Intermediate events represent the outcome of one or more events triggered by some events connected via a logical gate (e.g. AND, OR gates). They are not the top event and therefore these events themselves can act as input into logical gates located higher in the fault tree (closer to the top event).

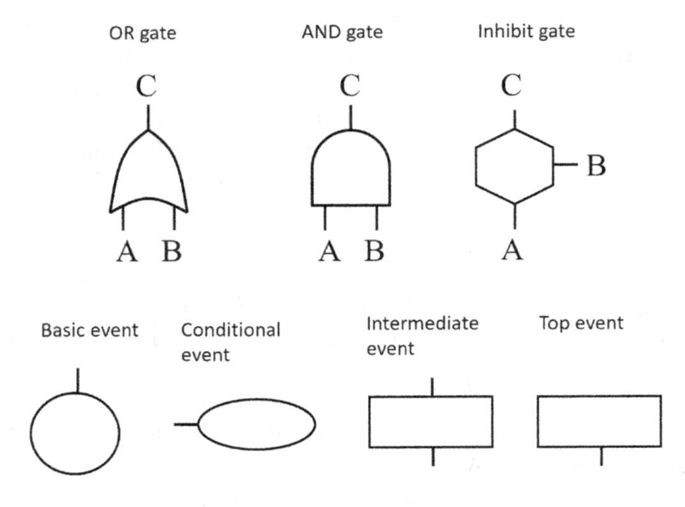

Figure 8.6 Graphical summary of basic building blocks for a fault tree.

8.6.1.3 Basic Events

Basic events are generally assumed to be statistically independent unless they are common cause failures. These are initiating events that require no further development, meaning no tree is drawn below these events.

8.6.1.4 Logical AND Gates

An AND gate is a digital logic gate that has two or more input signals and one output signal. The AND gate represents a situation where an output event occurs only if all input events have occurred. The order of the input events is not considered.

Consider an AND gate connecting two basic events A, B to and output event C as shown in Figure 8.6. Each of the events is described by a random variable that can take on two values: 0 if the event is not triggered, and 1 if the event is considered triggered:

$$A = \begin{cases} 1 & \text{event is triggered} \\ 0 & \text{event is not triggered} \end{cases}$$

$$B = \begin{cases} 1 & \text{event is triggered} \\ 0 & \text{event is not triggered} \end{cases}$$

$$C = \begin{cases} 1 & \text{event is triggered} \\ 0 & \text{event is not triggered} \end{cases}$$

The value of C is conditional on the values of A and B is summarised in a truth-table as shown in Table 8.1.

The probability of triggering the gate and evoking event C ($C = 1$) is therefore given as:

$$P(C=1) = P(A=1 \cap B=1) = P(A=1)P(B=1) \tag{8.15}$$

Assuming here that the input events A and B are statistically independent. The AND gate is similar to a parallel system consisting of two subsystems which only fails if both subsystems fail simultaneously (i.e. $A=1$ and $B=1$).

Table 8.1 Truth Table for AND Gate with Two Input Events

Event A	Event B	Event C
0	0	0
0	1	0
1	0	0
1	1	1

8.6.1.5 Logical OR Gates

The OR gate is a digital logic gate that has two or more input signals and one output signal. The OR gate represents situations where an output event is triggered if one or more of the input events have been triggered. The order of the input events is not considered. Consider an OR gate with the same two input events defined for the AND gate, i.e. A and B leading to an output event C. The corresponding truth-table for the OR gate is shown in Table 8.2.

For the OR gate the probability of triggering the output event C $(C = 1)$ is therefore given as:

$$\begin{aligned} P(C=1) &= P(A=1 \cup B=1) = P(A=1)+P(B=1)-P(A=1 \cap B=1) \\ &= P(A=1)+P(B=1)-P(A=1)P(B=1) \end{aligned} \tag{8.16}$$

As before, it is assumed that the input events A and B are statistically independent. As fault trees are generally applied to systems where the individual probabilities of failure are assumed very small, it is common to simplify Eq. 8.16 by assuming $P(A=1)P(B=1) \approx 0$, which reduces it to:

$$P(C=1) = P(A=1 \cup B=1) \approx P(A=1)+P(B=1) \tag{8.17}$$

This is a conservative estimate of the failure probability.

The OR gate is conceptually similar to a series system where a failure occurs if one or more of the subsystems fails, and the system survives only if none of the subsystems fails (i.e. $A=0$ and $B=0$).

8.6.1.6 Logical Inhibit Gate

The inhibit gate is similar to the AND gate but conceptually represents a transistor component from electronics, where a signal is only allowed to pass once an enabling event is activated. Consider again the two logical input variables A and B, where B is now considered an enabling event. This means the output event C can only be triggered $(C = 1)$ if event A is

Table 8.2 Truth Table for OR Gate with Two Input Events

Event A	Event B	Event C
0	0	0
0	1	1
1	0	1
1	1	1

triggered and event B is enabled (B = 1). In this case, the truth table for the inhibit gate is similar to that describing the AND gate in Table 8.1.

8.6.2 Construction of a Fault Tree

Once the functioning and interconnections of the entire system is understood, the construction of the fault tree can commence using the basic symbol listed in Figure 8.6.

8.6.2.1 *Minimal Cut Sets*

A cut set is a collection of basic events that will result in the triggering of the top event. There can be multiple cut sets for the same system. Of these cut sets, the minimal cut set is defined as the smallest combination of basic events which, if they occur, will cause the top event to occur. The complete collection of all minimal cut sets represents all the unique modes of system failure.

The MOCUS algorithm, developed by the US Atomic Energy Commission (USAEC; Fussell et al., 1974) can be used to identify the minimal cut set in any given fault tree composed of AND and OR gates. The algorithm works as follows:

1. Name each AND and OR gate in the tree from top down (e.g. G0, G1, G2, . . .).
2. Name each basic event in the tree (e.g. 1, 2, 3, . . .).
3. Create a matrix and add the first gate in the upper left cell (row 1, column 1).
4. Replace the gate number with the input events as follows:
    a. OR gate: replace gate number entry in the matrix by vertically filling the gate cell and cells below with the numbers of the basic events entering the gate.
    b. AND gate: replace the gate number entry in the matrix by horizontally filling the gate cell and cells to the right with the numbers of the basic events entering the gate.
5. Continue until all gates have been replaced by subsets of basic events.

At this stage, each row in the matrix will constitute a cut set, i.e. a set of basic events that can trigger the top event and cause a system failure. Finally, remove duplicate basic events within each cut set and remove any cut set that contains another but smaller cut set.

Consider the following example of a fault diagram reproduced from the USAEC (Fussell et al., 1974) report and shown in Figure 8.7. Initiating the MOCUS algorithm to find the minimum cut sets requires that all gates and basic events are labelled. Note that the same fundamental event (event 2) occurs in two locations, which is fine.

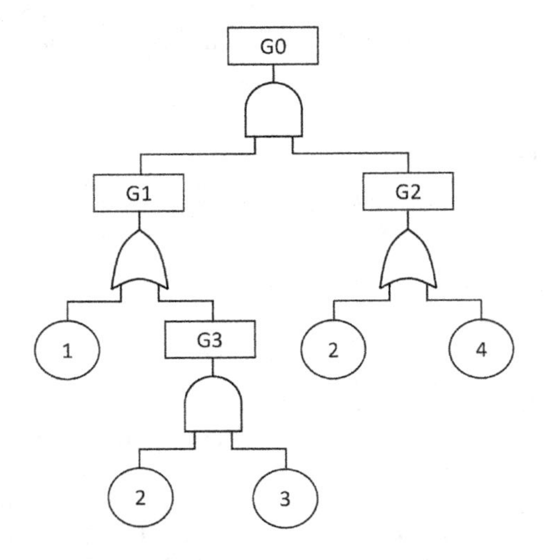

Figure 8.7 Example of fault tree consisting of two OR gates, one AND gate, four different basic events, and a top event. Gates are denoted G0 to G3, and the basic events denoted 1, 2, 3, and 4.

Steps 1 and 2: The fault diagram consists of a top event triggered by an AND gate (denoted G0), which itself is triggered by two events originating from OR gates (G1 and G2). The OR gate G1 is triggered by basic event 1 and an event triggered by an AND gate (G3) etc.

Step 3: Identify the top gate and put the gate name into the upper left cell of a matrix:

G0		

Step 4: Replace the gate name with the input events based on whether it is an AND (horizonal) or an OR (vertical) gate. As gate G0 is an AND gate, we replace the entry G0 with its input events (G1 and G2) in a horizonal manner and the matrix becomes:

G1	G2	

Next, gate G1 is an OR gate and is triggered by basic event 1 and the event related to gate G3. We therefore need to replace G1 vertically with 1 and G3 as:

1	G2	
G3	G2	

Next, gate G2 is an OR gate triggered by basic events 2 and 4, and the matrix therefore becomes:

1	2	
G3	2	
1	4	
G3	4	

Finally, both G3 gates in the first row of the matrix are resolved, noting that gate G3 is an AND gate triggered by basic events 2 and 3. Note that in each row the basic events have been ranked in ascending order.

1	2	
2	2	3
1	4	
2	4	3

As all gates have now been replaced with basic events, the algorithm has completed its work. The final matrix has four rows, and the algorithm has therefore identified four cut sets: {1,2}, {2,2,3}, {1,4}, and {2,4,3}. As the basic event 2 is repeated twice in the second cut set, the final cut set is reduced to {1,2}, {2,3}, {1,4}, and {2,4,3}. Three of the four cut sets consist of two events and are therefore all minimum cut sets: {1,2}, {2,3}, and {1,4}. This means that the system fails if any of these three minimum cut sets materialises.

Each of the minimum cut sets is triggered if both events occur. This is analogous to a parallel system where failure only occurs if all subsystems fail. In contrast, the overall system fails if one or more of the minimum cut sets are triggered, similar to a series system. We can therefore consider the system to be a series system where each individual subsystem is a minimum cut set as shown in Figure 8.8.

Finally, the probability of failure of the complex system shown in Figure 8.8 can be calculated as:

$$P_f = 1 - \prod_{i=1}^{n} (1 - PC_i) \qquad (8.18)$$

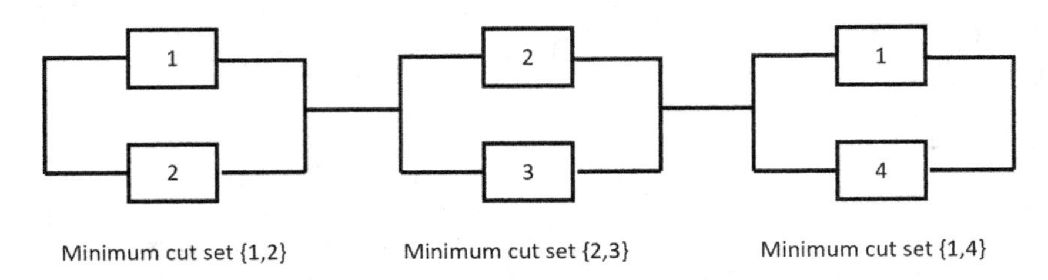

Figure 8.8 Series system consisting of three parallel subsystems, each representing a minimum cut set.

Where n is the number of minimum cut sets and PC_i is the probability of failure of each cut set. Consider the system in Figure 8.8 and denote the probability of each basic event as P_1, P_2, P_3, and P_4, and the three minimum cut sets {1,2}, {2,3}, and {1,4}, then the total probability of failure of the system is:

$$P_f = 1 - (1 - PC_1)(1 - PC_2)(1 - PC_3) \tag{8.19}$$

Where the probability of each cut set is defined according to the probability of triggering the basic events as:

$$\{1,2\}: \; PC_1 = P_1 \times P_2$$

$$\{2,3\}: \; PC_2 = P_2 \times P_3$$

$$\{1,4\}: \; PC_3 = P_1 \times P_4$$

As each minimum cut set is synonymous with an additional subsystem in a parallel system, it follows that we should aim to reduce the number of minimum cut sets.

REFERENCE

Fussell, J. B., Henry, E. B. and Marshall, N. H. (1974). *MOCUS: A computer program to obtain minimal sets from fault trees (No. ANCR-1156)*. Idaho Falls, ID, USA: Aerojet Nuclear Co.

Chapter 9

Decision-Making in an Uncertain World

9.1 A BASIC DECISION-MAKING MODEL

Decision-making is a fascinating topic spanning several scientific disciplines including economics, business, mathematics, and psychology. In this chapter only a brief introduction is provided, but interested readers are encouraged to produced further reading on more advanced topics in the rich existing literature.

A decision-making problem can be formulated as consisting of the following three components: Actions, States of Nature, and Payoff as introduced in the following.

Actions (a_i): A single decision-maker has a choice between different Actions (or decision) that are available. An exhaustive list of Actions (all available decisions) consists of n different Actions as follows:

$$a_i, i = 1, \ldots, n \tag{9.1}$$

The decision-maker must take one of the Actions but can only take *one* Action.

States of Nature (s_j): Deciding on the "best" Action is easy if there is complete knowledge of the external factors influencing the decision. However, most practical cases are more complex, and an Action should be chosen based on a series of possible external conditions outside the control of the decision-maker. In the decision-making literature such external factors are named *States of Nature*. An exhaustive list of States of Nature consists of k different states as given below:

$$s_j, j = 1, \ldots, k \tag{9.2}$$

The decision problem now involves choosing the *best* Action a_i not knowing the State of Nature, s_j.

DOI: 10.1201/9781032700373-9

Payoff (π_{ij}): The combination of a State of Nature and an Action will result in Payoff, often measured in monetary terms, which can be a profit or a loss (positive or negative):

$$\pi_{ij} = \pi\left(a_i, s_j\right) \text{for } i = 1,\ldots n \text{ and } j = 1,\ldots k \tag{9.3}$$

Think of π_{ij} as the consequence of taking Action a_i under State of Nature s_j. Calculation of π_{ij} can be challenging as it involves all monetary impacts of each State of Nature and each Action. The Payoffs can be organised in a Payoff Table, where the *k* columns represent different States of Nature, and the *n* rows represent Actions available to the decision-maker.

From the Payoff Table it is also possible to define an equivalent Opportunity Loss Table. The opportunity loss is incurred by having selected a sub-optimal Action for a given State of Nature. The elements of this table are calculated by considering each State of Nature in turn. The Opportunity Loss, $o(a_i, s_j)$, is calculated by subtracting the Payoff for each Action, π_{ij}, from the maximum possible Payoff for this State of Nature, $max(\pi_{ij})$. Mathematically this is defined as:

$$o\left(a_i, s_j\right) = \max_i\left(\pi_{ij}\right) - \pi_{ij} \tag{9.4}$$

As the exact State of Nature is unknown, the opportunity loss arises because of uncertainty from factors that are not under the control of the decision-maker. The optimal opportunity loss is zero, in which case the chosen Action is the best possible Action for a given State of Nature.

EXAMPLE 9.1 THE PAYOFF TABLE

A construction company is planning to build a new housing estate. They are trying to decide the number of dwellings to construct to optimise profit, and consider three possible sizes of the new estate: a_1 a small estate, a_2 a medium size estate, or a_3 a large estate. These three options are the Actions available to the decision-maker. However, the company is not sure about the future demand for housing and the price they can request. They consider three possible future States of Nature: s_1 low demand, s_2 medium demand, and s_3 high demand. The anticipated profit, in tens of thousands of pounds, for each combination of estate size (a_1, a_2, and a_3) and future demand (s_1, s_2, and s_3) are shown in the Payoff Table below. The payoffs reflect that, for example, low demand combined with a large estate will result in a smaller profit margin (see the table below).

The Payoff Table for the house builder decision problem is as follows:

	s_1	s_2	s_3
a_1	39	50	57
a_2	32	60	75
a_3	20	53	88

A Payoff Table is commonly constructed so that columns represent States of Nature and rows represent Actions. What Action should the company take in this example? If the future demand was known with absolute certainty, then it would be trivial to select the optimal strategy.

1. s_1 low demand →, choose Action a_1 small estate.
2. s_2 medium demand →, choose Action a_2 medium estate.
3. s_3 high demand →, choose Action a_3 large estate.

However, the future State of Nature is not known with certainty which makes the decision problem more difficult.

The resulting Opportunity Loss Table is calculated using Eq. 9.4.

Opportunity loss table for house building decision problem and derived from Payoff table above is presented in the following:

	s_1	s_2	s_3
a_1	39–39 = 0	60–50 = 10	88–57 = 31
a_2	39–32 = 7	60–60 = 0	88–75 = 13
a_3	39–20 = 19	60–53 = 7	88–88 = 0

An opportunity loss of zero (no regret) occurs when the optimal Action is chosen for a particular State of Nature.

9.2 WHAT IS A "GOOD" DECISION?

Deciding what Action to take given that the exact future State of Nature is uncertain is a subjective choice, reflecting the decision-maker's aversion to risk. Several different strategies can be defined, each representing a particular attitude to risk from the perspective of the decision-maker. Some of the common strategies are introduced below. In these simple cases, all States of Nature are assumed to be equally likely to occur.

Maximax: This is the strategy of choice for an optimist. The decision-maker considers the Payoff available for each Action, and simply chooses the Action that includes the highest Payoff across all states of nature.

Maximin: This is the strategy that a pessimist would adopt. This strategy assumes that whatever Action is chosen, the worst possible outcome will materialise (minimum Payoff). The strategy is therefore to choose the Action with the highest (maximum) minimum Payoff.

Minimax: The Minimax strategy minimises the maximum opportunity loss (or regret). First, the maximum opportunity loss is identified for each State of Nature. Next, the maximum opportunity loss associated with each Action is identified (boxed). Finally, the Action with the minimum maximum opportunity loss is chosen.

Principle of insufficient reason: This strategy is aimed at a decision-maker who is neither pessimistic nor optimistic and assumes all States of Nature are equally likely. The decision criterion is derived by calculating the sum of all possible Payoffs for each Action, and selecting the Action with the largest sum as formulated below:

$$\max_i \left(\sum_{j=1}^{k} \pi(a_i, s_j) \right), i = 1,\ldots,n \tag{9.5}$$

Later we shall see how Eq. 9.5 can be enhanced to also consider cases where some States of Nature are considered more likely than others.

EXAMPLE 9.2 ILLUSTRATION OF DECISION-MAKING STRATEGIES

Consider the Payoff Table defined in the previous example. Here, we use this table to make decisions using the different decision-making strategies as follows:

Maximax: From the Payoff Table, identify the maximum Payoff possible under each Action (grey), and select the Action with the maximum Payoff (a_3). The table below shows the Maximax table:

	s_1	s_2	s_3	*Max Max* π
a_1	39	50	57	57
a_2	32	60	75	75
a_3	20	53	88	88

Maximin: From the Payoff Table, identify the minimum Payoff associated with each Action (boxed), and select the Action resulting in the maximum minimum Payoff (a_1).

	s_1	s_2	s_3	*Max Min* π
a_1	[39]	50	57	[39]
a_2	[32]	60	75	32
a_3	[20]	53	88	20

Minimax: The Minimax criterion is applied by choosing the Action with the minimum opportunity loss across all States of Nature. First identify the maximum opportunity loss associated with each Action (boxed), and then selected the Action associated with the minimum maximum opportunity loss (a_2). The following table presents the Minimax table:

	s_1	s_2	s_3	*Min Max o*
a_1	39–39 = 0	60–50 = 10	88–57 = [31]	31
a_2	39–32 = 7	60–60 = 0	88–75 = [13]	[13]
a_3	39–20 = [19]	60–53 = 7	88–88 = 0	19

Principle of Insufficient Reason (PIR): For each Action, sum up the Payoff across all States of Nature, and select the Action with the highest total Payoff (a_2). The PIR table is presented below:

	s_1	s_2	s_3	$\sum \pi$
a_1	39	50	57	39 + 50 + 57 = 146
a_2	32	60	75	32 + 60 + 75 = [167]
a_3	20	53	88	20 + 53 + 88 = 161

Looking at the above four options, it can be noticed that each strategy results in a different outcome depending on the decision-maker's attitude to risk, whether optimistic, pessimistic, or neutral.

9.3 DECISION-MAKING WITH PROBABILITY

The strategies listed in the previous section do not consider any information regarding the probability of the different States of Nature, so effectively ignore any information as to whether some futures are more likely than others. However, in some cases a probability can be assigned to the States of Nature, e.g., from statistical analysis of data or expert opinion, which can help improve the decision-making process.

Consider the case where each State of Nature, $s_j, j = 1,\ldots,k$, is associated with a probability $P(s_j) = p_j, \geq 0$ for all j, and the sum of all p_j equals one, i.e., $\sum_{i=1}^{k} p_j = 1$. In this case, the Expected Value (EV) of Action $a_i, i = 1,\ldots,n$ is defined as:

$$EV(a_i) = \sum_{j=1}^{k} \pi(a_i, s_j) P(s_j) \tag{9.6}$$

A rationale strategy would be to choose the Action among the $i = 1,\ldots,n$ alternatives that maximises the expected value, as formulated below:

$$\max_i (EV(a_i)),\ i = 1,\ldots,n \tag{9.7}$$

EXAMPLE 9.3 MAKE DECISIONS BASED ON *EVA*

Assume that the construction company introduced in Example 9.1 has assigned the following probabilities to the three possible future States of Nature: probability of low demand $P(s_1) = 0.40$, probability of medium demand $P(s_2) = 0.40$, and probability of high demand $P(s_3) = 0.20$. Calculate the Expected Value of Payoff for each Action using Eq. 9.7 and select the Action with the maximum $EV(a_2)$.

	$s_1(0.40)$	$s_2(0.40)$	$s_3(0.20)$	EV
a_1	39	50	57	39 × 0.40 + 50 × 0.40 + 57 × 0.20 = 47.0
a_2	32	60	75	32 × 0.40 + 60 × 0.40 + 75 × 0.20 = **51.8**
a_3	20	53	88	20 × 0.40 + 53 × 0.40 + 88 × 0.20 = 46.8

Note that if Action a_2 is chosen, then the actual Payoff for this decision will not be 51.8 on this one occasion. It will be either 32, 60, or 75 with a probability 0.40, 0.40, and 0.20, respectively. The same way that when a six-sided die is rolled once, the outcome is either 1, 2, 3, 4, 5, or 6. But if the same die is rolled a very large number of times then the average number of pips is 3.5 $(\text{i.e., } 1/6 \times (1+2+3+4+5+6) = 3.5)$. Interestingly, probability theory cannot predict the outcome of the next event, but it can tell us about the long-term behaviour of the system. In the same way, if we had to take Action a_2 in the same decision problem many times, then on average the outcome would be a payoff of 51.8.

9.4 DECISION TREES

Sometimes it can be useful to illustrate the decision-making process using a decision tree. A decision tree is a graphical representation of the decision-making problem consisting of nodes and branches representing different aspects of the decision-making process as it progresses over time. There are two types of nodes: a decision node (a square, □) controlled by the decision-maker, and a chance node (a circle, ○) which is outside the control of the decision-maker. Moving from left to right, branches leaving a decision node represent Actions, while branches leaving a chance node represent States of Nature. At the end of each branch on the right-hand side the Payoff (π_{ij}) corresponding to the ith Action and the jth State of Nature is noted.

An example of a decision-tree is shown in Example 9.4. Once the decision tree has been constructed, the decision-problem is solved by moving from right to left following these three steps:

1. When encountering a chance node, calculate the Expected Value (EV) corresponding to the States of Nature to the right of the node. When completed, remove the tree to the right of the node.
2. At each decision node, select the Action corresponding to the highest EV. When completed replace all of the tree to the right of the node with this EV.
3. Repeat steps 1 and 2 until the start node is reached.

EXAMPLE 9.4 DECISION TREE FOR HOUSE BUILDING EXAMPLE

The decision tree shown in the figure below illustrates the initial setup, including all the components of the basic decision-making problem: chance nodes, decision nodes, probabilities, Actions, States of Nature, and Payoffs.

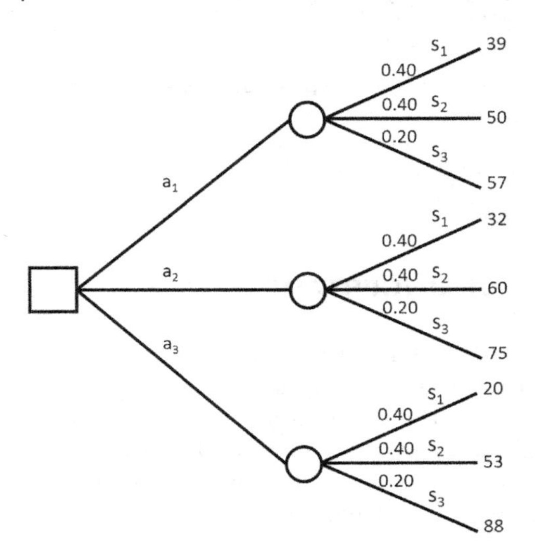

Figure A Decision tree, setup

The analysis of the decision tree starts with step 1, calculating the *EV* corresponding to the States of Nature to the right of the chance nodes ○ resulting in *EVs* of 47.0, 51.8, and 46.8. In step 2 the Action associated with the maximum *EV* value is taken, and all other Actions dismissed (crossed out). In this example, the Action with the highest EV is a_2. The completed decision tree is shown in the figure below:

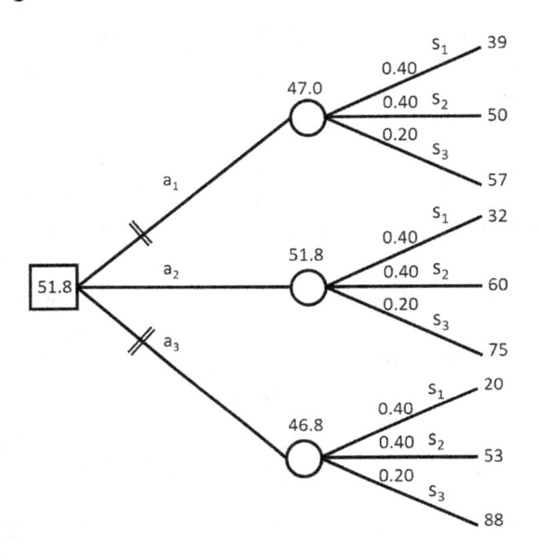

Figure B Decision tree, completed.

As the final decision node □ has been reached, the decision tree is complete. It is possible to construct decision trees for much more complex cases than illustrated here, in which case the procedure is similar but continues until the final decision node is reached.

9.5 THE COST OF UNCERTAINTY

If the State of Nature is known with absolute certainty, then it is straightforward to select the strategy that maximises the Payoff. What is the value of absolute certainty to the decision-maker? Recall from section 9.2 that the maximum Payoff for a particular State of Nature s_j is found by considering the Payoff as sociated with all $i = 1,\ldots,n$ Actions for s_j, i.e., the maximum value in each column in the Payoff Table:

$$\pi^*(s_j) = \max_i(\pi_{ij}),\ i = 1,\ldots,n \tag{9.8}$$

In the case where the decision-maker can always select the most profitable payoff the Expected Value under certainty ($EVUC$) is defined as:

$$EVUC = \sum_{j=1}^{k} \pi^{*}(s_j) P(s_j) \tag{9.9}$$

Next, the Expected Value of perfect information ($EVPI$) evaluates the value of perfect knowledge, i.e., moving from EVA (uncertain) to EVUC (certain). EVPI is therefore defined as the difference between the $EVUC$ and the expected value (EV) derived in Eq. 9.6, i.e.:

$$EVPI = EVUC - EV \tag{9.10}$$

If a decision-maker is given the option of acquiring perfect knowledge about the State of Nature for less than the value of perfect information (and it is legal to do so), then that could be a profitable strategy.

EXAMPLE 9.5 COST OF UNCERTAINTY FOR HOUSE BUILDER

If the decision-makers of the construction company have perfect knowledge of the future States of Nature (low, medium or high demand for housing), they can take the Action that guarantees maximum profit in each case. Use Eq. 9.9 to calculate the maximum Payoff for each State of Nature identified the boxed values in the Payoff Table below:

	$s_1(0.40)$	$s_2(0.40)$	$s_3(0.20)$
a_1	$\boxed{39}$	50	57
a_2	32	$\boxed{60}$	75
a_3	20	53	$\boxed{88}$

Next, the expected value under certainty, $EVUC$, can be calculated using Eq. 9.9 as:

$$EVUC = 39 \times 0.40 + 60 \times 0.40 + 88 \times 0.20 = 57.2$$

and therefore, the expected value of perfect information becomes:

$$EVPI = 57.2 - 51.8 = 5.4$$

This example shows if the company can acquire perfect information for less than 5.4, then that is a worthwhile strategy.

9.6 EXPECTED UTILITY

Using the expected value of the Payoff as a decision variable might overlook an important aspect of the problem which is the decision-maker's attitude to the prospect of earning or losing money. The concept of expected utility has been introduced by economists as a way of representing a decision-maker's attitude to risk.

The key concept in expected utility theory is the utility function, $U(w)$, which measures a decision-maker's utility (or well-being, or satisfaction) derived from a certain amount of wealth w. Utility does not have a physical unit, but represents a measure of satisfaction, benefit, or well-being associated with a particular outcome. Consider two possible outcomes of a future Action (w_1 and w_2) and note that a decision-maker prefers w_1 over w_2. In that case, the utility of w_1 is larger than the utility of w_2, i.e., $U(w_1) > U(w_2)$.

Just as the Expected Value discussed in section 9.3, the Expected Utility can be derived as follows:

$$EU(a_i) = \sum_{j=1}^{k} U(w_{ij}) P(s_j) \tag{9.11}$$

where w_{ij} is the wealth resulting from Action i and State of Nature j, and $P(s_j)$ is the probability of State of Nature j. Similar to the expected value in Eq. 9.6, a rationale decision-maker would choose the Action among the $i = 1,\ldots,n$ alternatives that maximises the expected utility, as formulated below:

$$\max_i (EU(a_i)), i = 1,\ldots,n \tag{9.12}$$

The form of the utility function $U(w)$ reflects the decision-maker's attitude to risk. For a risk-averse person, it is more important to maintain a certain level of wealth than to gamble for marginal gains. A risk lover will take-on high-stakes gambles with the promise of a big payout (think, for example, a bank robber), while for a neutral approach consideration is not influenced by the potential gains or losses. Examples of utility curves for risk-averse $U(w) = \sqrt{(w)}$, risk lover $U(w) = w^2$, and risk neutral $U(w) = w$ decision-makers are shown in Figure 9.1.

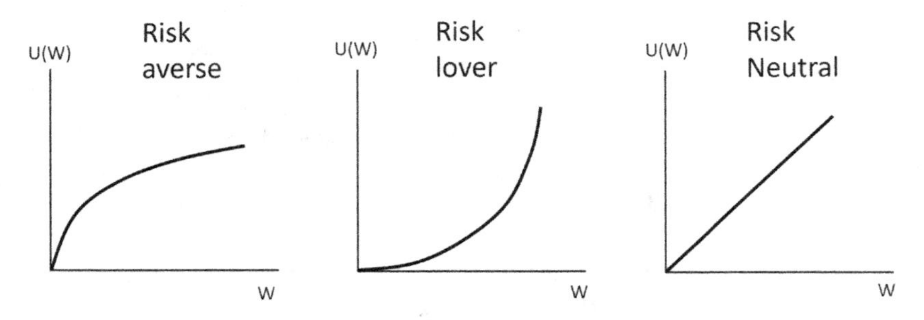

Figure 9.1 Examples of utility curves for a decision-maker who is risk averse (*left*), a risk lover (*centre*), and risk neutral (*right*).

EXAMPLE 9.6 DECISION-MAKING BASED ON EXPECTED UTILITY

A housing company has the opportunity to buy a plot of land at a prime location for developing a future housing project. The company has a current wealth of $w = £1{,}200{,}000$ available and the cost of the land is $£1{,}000{,}000$. The development of the land will potentially increase the company's total wealth to $w = £10{,}200{,}000$. The only problem is that the land does not have planning permission. Experts estimate that in the face of local opposition to development, the probability of gaining planning permission is $p = 0.10$. If planning permission is not obtained, then the company estimate that they can sell it back for £50,000 leaving a total wealth of £250,000. Should the company invest in the land or not?

The decision problem is organised in the table below, all numbers quoted in million £.

	PP granted (p)	*PP rejected* $(1-p)$
a_1 Buy	10.20	0.25
a_2 Do not buy	1.20	1.20

First, calculate the Expected Value (in GBP millions) of each investment decision (buy/not buy land)

$$EV(a_1) = 10.20 \times 0.10 + 0.25 \times (1 - 0.10) = 1.25$$

$$EV(a_2) = 1.20 \times 0.10 + 1.20 \times 1.20 = 1.20$$

Therefore, from the perspective of the EV, the optimal Action is a_1 as it yields the highest expected income. However, the company might be reluctant to make this gamble as there is a high risk of losing money and only a small chance of winning (but winning big). If the decision-maker is risk-averse and has a utility function expressed as $U(w) = \sqrt{(w)}$, then the expected utility according to Eq. 9.11 is:

$$EU(a_1) = U(10.20) \times 0.10 + U(0.25) \times (1 - 0.10) = 0.77$$

$$EU(a_2) = U(1.20) \times 0.10 + U(1.20) \times (1 - 0.10) = 1.10$$

As the expected utility of a_2 is higher than the expected utility for a_1, the company will choose a_2.

Question: How high should the probability of getting planning permission be before the company goes ahead with the land purchase?

Answer: So high that the expected utility of a_1 exceeds the expected utility of a_2, i.e. solving $EU(a_1) > EU(a_2)$ with the probability p as the unknown:

$$U(10.2) \times p + U(0.25) \times (1 - p) > U(1.2)$$

$$p > \frac{U(1.2) - U(0.25)}{U(10.2) - U(0.25)} = 0.22$$

Chapter 10

Hypothesis Testing

10.1 INTRODUCTION

Hypothesis testing is a statistical procedure used to examine a proposed explanation to a particular phenomenon (a hypothesis). More formally, the technique allows probabilistic statements to be made about parameters of probability distributions based on sample data. For example, how confident can we be that the strength of a material exceeds a certain value based on a set of laboratory tests? To get a better idea of how hypothesis testing might be used in practice, consider the following example. A construction material company has developed a new procedure for using fly-ash–based polymer concrete (FAPC) as a replacement for ordinary Portland cement concrete (OPCC). Results from laboratory testing of the compressive strength involving 25 samples of the new products are summarised in Figure 10.1, along with the sample mean and standard deviation of the 25 test results.

The box plot in Figure 10.1 shows considerable variation between the 25 test results. The company is now interested in knowing if the strength of the new product is different from the known value of OPCC. Assuming that the strength of FAPC is a random variable with a probability distribution, then the hypothesis testing in this example will examine if there really is a statistical difference between the strength of FAPC and the known strength of OPCC.

In the next sections, the most important aspects of hypothesis testing will be introduced, followed by examples of application to specific problems often encountered in civil engineering data analysis.

10.2 SAMPLE VERSUS POPULATION

Before introducing hypothesis testing, it is important to highlight the difference between populations and samples. This topic was covered in Chapter 3 when discussing parameter estimation, but it is key to understanding ideas about hypothesis testing and therefore is further expanded here.

DOI: 10.1201/9781032700373-10

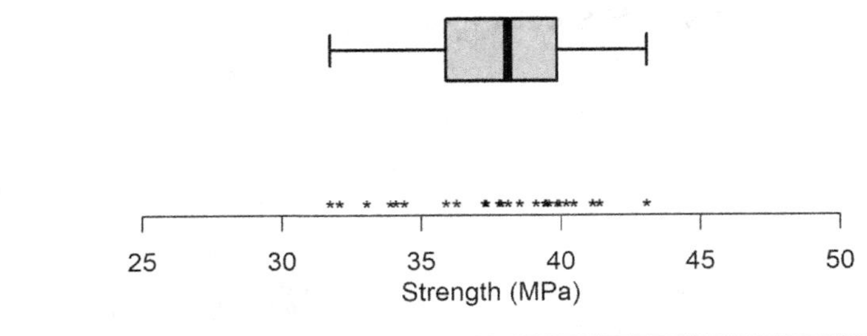

Method	Mean (MPa)	Standard deviation (MPa)
FAPC	37.8	2.7

Figure 10.1 Box plots comparing compressive strength (MPa) of 25 samples of fly-ash–based polymer concrete. The asterisks indicate the individual observations. The table shows the sample mean and standard deviation of the 25 experimental results.

In the example above, 25 tests were conducted resulting in a sample consisting of 25 values, each representing the strength of FAPC. If the strength of the concrete is described by a random variable, then automatically it is also assumed that there is a mean value and a standard deviation. Before the experiments were conducted the actual values of these parameters were unknown, but once the 25 data points are available then estimates of these values can be obtained using the sample estimators in Eq. 3.1. Thus, it is important to distinguish between what is called the *population parameters*, which represent the true but unknown values, and the *estimates* of the *population* parameters obtained from a sample.

To illustrate the difference between population and sample values, consider again the strength of FAPC.

Assume that the strength of FAPC is a random variable X following a normal distribution with a population mean value of 36.5 MPa and a standard deviation of 4.0 MPa, i.e.:

$$X \sim N(36.5, 4.0^2) \tag{10.1}$$

In a practical setting, these population values are not known, but each of the 25 tests will result in a realisation of the random variable X. Using the Monte Carlo simulation technique described in Chapter 6, it is possible to generate 25 realisations of X based on the population values. Figure 10.2 shows three examples of 25 random realisations of X. For each example, the sample mean and standard deviation are calculated (using Eq. 3.1). Observe how these sample values are close to each other (and close to the population parameters specified in Eq. 10.1) but slightly different in the three figures. This difference is important and represents the sampling variability.

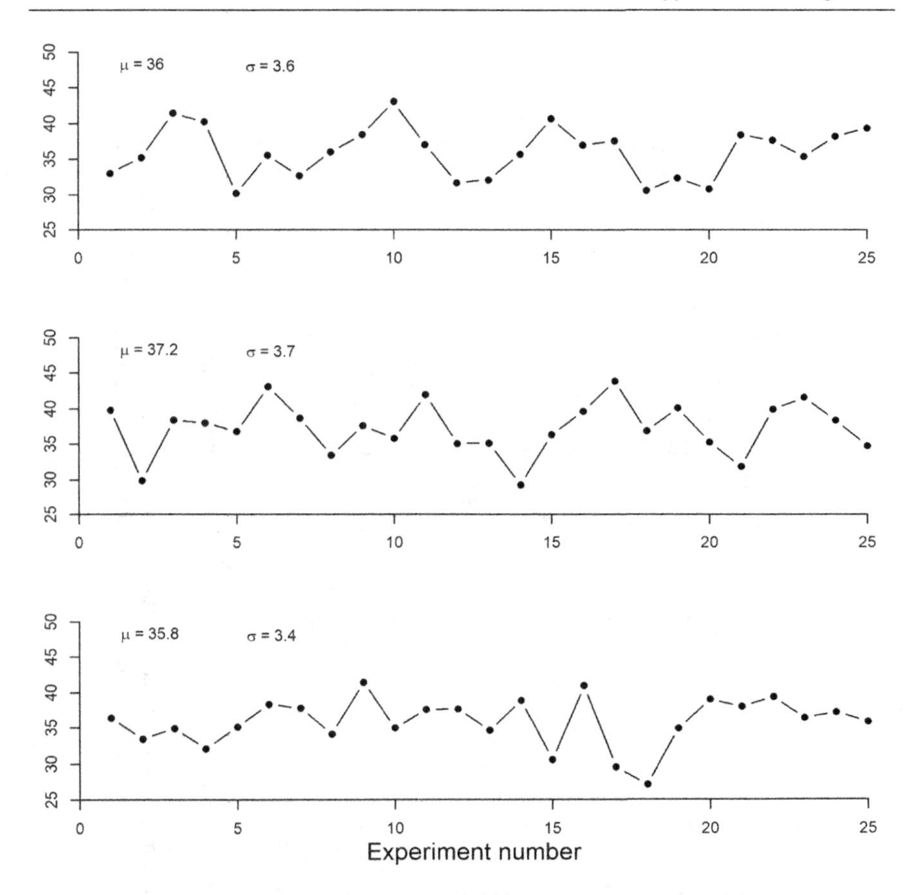

Figure 10.2 Three samples each containing 25 realisations generated by a Monte Carlo simulation based on the normal distribution defined in Eq. 10.1.

In Figure 10.3 the same experiment has been replicated 1000 times, and for each experiment the sample mean and standard deviation of the 25 realisations has been calculated. The two histograms in Figure 10.3 show that the estimated sample mean values and standard deviations obtained from the 1000 experiments can vary across a range of values but are centred around the population value of 36.5 MPa and 4.0 MPa as specified in Eq. 10.1.

The takeaway from this section is that a sample (a set of data points) represents only a subset of all possible values, and therefore estimates of statistical parameters obtained from a sample represent a best guess of the population values, and this best guess is itself uncertain. Therefore when analysing a dataset, always be mindful that it likely represents only a small subset of all possible values.

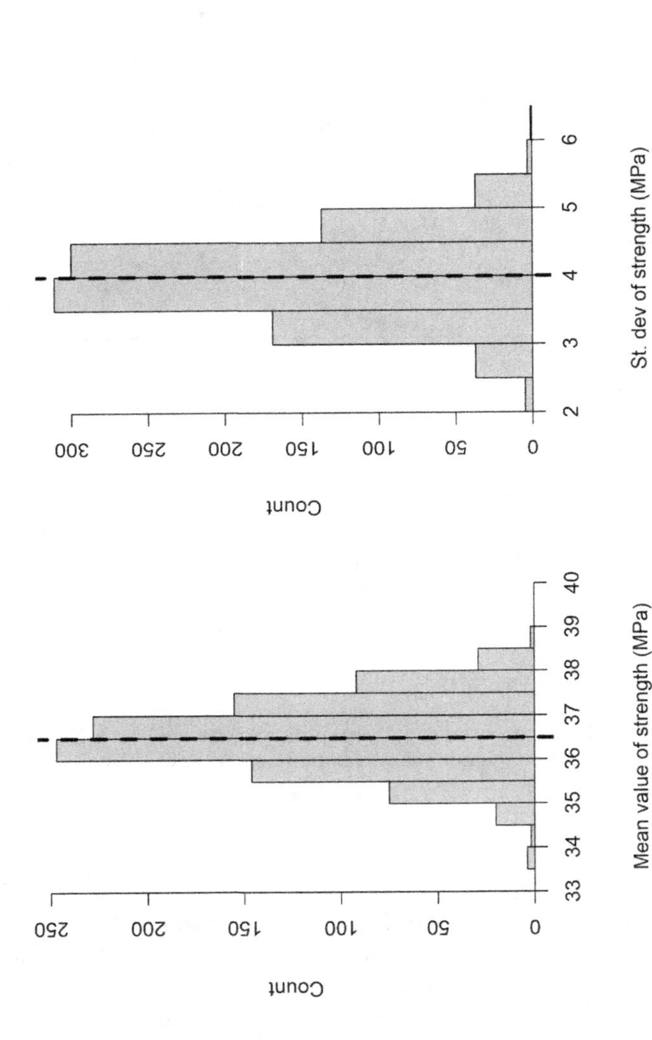

Figure 10.3 Histograms of counts summarising (*left*) sample mean and (*right*) sample standard deviation obtained from 1000 replicated experiments involving 25 tests. The thick vertical hatched lines represent the population mean and standard deviation specified in Eq. 10.1, i.e. 36.0 MPa and 4.0 MPa.

10.3 THE CENTRAL LIMIT THEOREM

The ideas of sampling presented in the previous section can be formalised via the central limit theorem (CLT). Consider a sample containing a total of n independent random variables X_i drawn from a population with a mean value μ and standard deviation σ, and let $\bar{X}_n$ be the sample mean value calculated as:

$$\bar{X}_n = \frac{1}{n}\sum_{i=1}^{n} X_i \tag{10.2}$$

Note here that $\bar{X}_n$ is a function of the n random variables and therefore itself a random variable. The central limit theorem states that as the sample size n increases (goes towards infinity), the probability distribution of $\bar{X}_n$ converges towards a normal distribution as:

$$\lim_{n\to\infty} P\left(\frac{\bar{X}_n - \mu}{\sigma / \sqrt{n}} \le x\right) = \Phi(x) \tag{10.3}$$

This means that for large samples ($n > 30$ generally acceptable) the sample mean value is approximately normally distributed regardless of the distribution of the underlying population. An easier notation might be:

$$\bar{X}_n \sim N\left(\mu, \frac{\sigma^2}{n}\right) \tag{10.4}$$

The important aspects to note are that (1) this is true regardless of the form of the underlying distribution of the data (as long as the sample is large enough), and (2) the standard deviation of the mean value is smaller (by a factor of $1/\sqrt{n}$) than the standard deviation of the individual observations. Returning to the example above, the sample mean of the 25 FAPC strength tests shown in Figure 10.1 would be approximately normally distributed with a with a mean value of 37.8 MPa and a standard deviation of $2.7/\sqrt{25}$, i.e.:

$$\bar{X}_{25} \sim N\left(37.8, \frac{2.7^2}{25}\right) \tag{10.5}$$

The important takeaway is that regardless of the distribution of the underlying sample, the normal distribution is a good approximation to the mean value of the sample, even for modest sample sizes. This is an important insight for the ensuing discission of hypothesis testing.

10.4 WHAT IS A STATISTICAL HYPOTHESIS?

The first concept in hypothesis testing that needs introducing is the actual hypothesis that is going to be tested. The hypothesis typically represents a statement about a phenomenon, for example, the strength of FAPC concrete is similar to that of OPCC, as per the example above. This statement is called the null hypothesis and is denoted H_0. To allow for a statistical

EXAMPLE 10.1 THE CENTRAL LIMIT THEOREM AND THE UNIFORM DISTRIBUTION

Consider a random variable X that follows a uniform distribution as:

$$X \sim U(0,1)$$

This implies, with reference to Eqs. 2.31 and 2.32 that X has a mean value and standard deviation of:

$$\mu = E(X) = \frac{1+0}{2} = \frac{1}{2}$$

$$\sigma^2 = V(X) = \frac{(1-0)^2}{12} = \frac{1}{12}$$

According to the central limit theorem, Eq. 10.4, this implies that the sample mean $\bar{X}_n$ is normally distributed as:

$$\bar{X}_n \sim N\left(\frac{1}{2}, \frac{1}{12n}\right)$$

To illustrate how effective the central limit theory works, a set of Monte Carlo experiments is conducted using the principles set out in Chapter 6 for generating random numbers. In this example four different sample lengths are considered: $n=1$, $n=5$, $n=25$, and $n=100$ data points. For each of the four record lengths, the Monte Carlo experiment will generate 10,000 random samples from the uniform distribution describing the individual events and summarise each individual sample by calculating the mean value. Finally, a histogram is drawn of the 10,000 values and visually compared to the normal distribution of the mean value according to Eq. 10.4 (CLT); the results are shown in the figure below. Each of the plots shows the histogram of the 10,000 values of the sample mean and (1) the original uniform distribution of the individual observations (the horizontal line at density of 1) and (2) the approximate normal distribution of the mean value as per Eq. 10.4.

The first plot shows that when the sample length is $n=1$ (i.e. a single observation) then the mean value is simply the value itself (per Eq. 10.4), and thus the distribution of the mean will be identical to the underlying uniform distribution and nothing like a normal distribution. However, as the sample size increases the histogram of the mean values quickly starts to mimic that of a normal distribution, even for a small sample length of $n=5$. For very large samples $(n=100)$, there is a very good match between the histogram and the pdf of the normal distribution defined according to the central limit theorem. Also, note that the more data available for estimation of the mean (larger values of n), the smaller the standard deviation of the estimated mean.

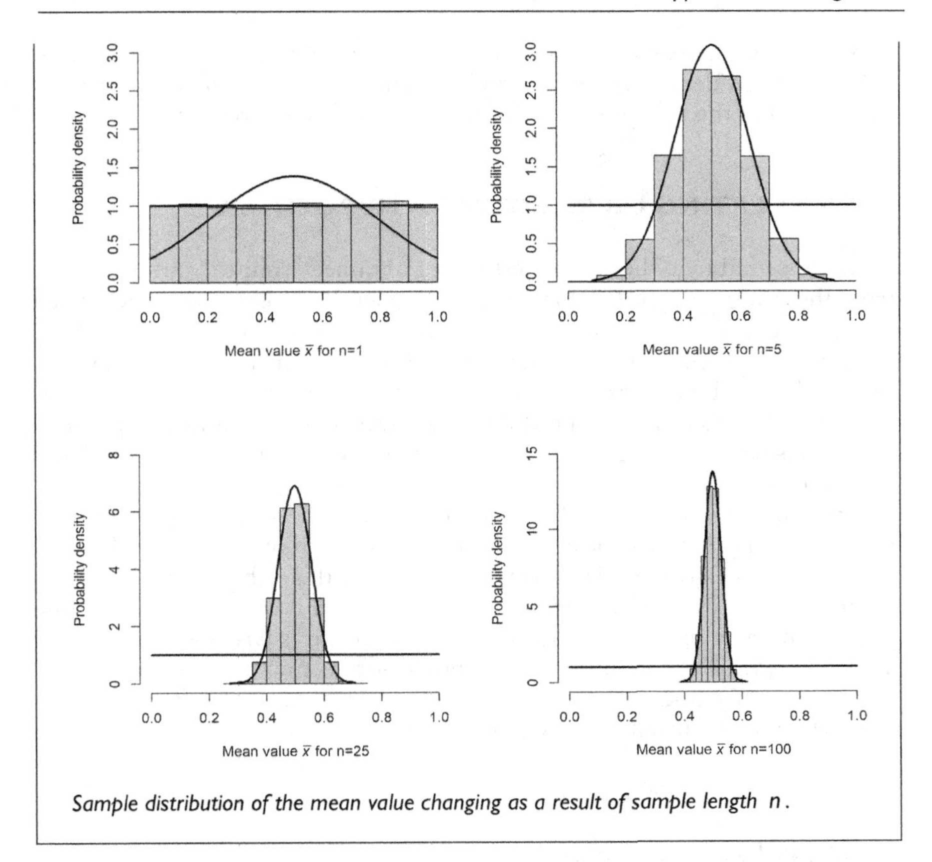

Sample distribution of the mean value changing as a result of sample length n.

treatment, the hypothesis needs to be translated into a statement about the parameters of the distribution describing the strength of FAPC. For example, the null hypothesis H_0 is that the FAPC is no different from OPCC, and therefore the "true" population mean value of the strength of FAPC, μ, should be equal to the equivalent mean value of OPCC, μ_0. In mathematical terms, the null hypothesis H_0 can be defined here as:

$$H_0 : \mu = \mu_0$$

The alternative to this hypothesis is that the strengths of the two concrete types are not equivalent. This is called the alternative hypothesis and is denoted H_1:

$$H_1 : \mu \neq \mu_0$$

The idea of hypothesis testing is that the null hypothesis can be either accepted or rejected. If the null hypothesis is rejected, then the alternative hypothesis is accepted. Generally, the alternative hypothesis represents what is hoped to be demonstrated (or find evidence for) in the analysis – in this example, that the two types of concrete have different strengths.

10.5 ACCEPTING OR REJECTING A HYPOTHESIS

As discussed above, the sample estimates obtained from the data can vary from the true population values, and consequently decisions as to accepting or rejecting the null hypothesis might be correct or incorrect. It is possible to make two different types of errors when drawing conclusions from experimental data; a Type I error represents the case of rejecting a null hypothesis even if it is true, while a Type II error involves accepting a null hypothesis even if it is not true (i.e. false). Table 10.1 shows the definitions of Type I and Type II errors.

From the 25 experiments listed in Figure 10.1 the sample mean value $\bar{x}$ has been derived, representing the estimated strength of FAPC. The decision to either accept or reject H_0 should be based on this value of $\bar{x}$. If $\bar{x}$ is too far from μ_0 then we should reject H_0 and vice versa if the value of $\bar{x}$ is close to μ_0. Thus, $\bar{x}$ is the criterion on which the decision is based and is denoted a test statistic. The condition for accepting or rejecting H_0 can now be formulated mathematically by considering an interval around μ_0 where values of c define the width of the interval as:

$$\text{Accept } H_0 : \mu_0 - c \leq \bar{x} \leq \mu_0 + c$$

$$\text{Reject } H_0 : \bar{x} < \mu_0 - c \cup \bar{x} > \mu_0 + c$$

The region of values for which $\bar{x}$ will result in H_0 being rejected is called the *critical interval*. Assuming in this example that material strength is normally distributed, and the null hypothesis is true (i.e. $\mu = \mu_0$), then the situation is illustrated in Figure 10.4.

Table 10.1 Definitions of Type I and Type II Errors

	True Situation (Population Parameters)	
Decision	H_0 *is True*	H_0 *is False*
Reject H_0	Type I error	Correct decision
Accept H_0	Correct decision	Type II error

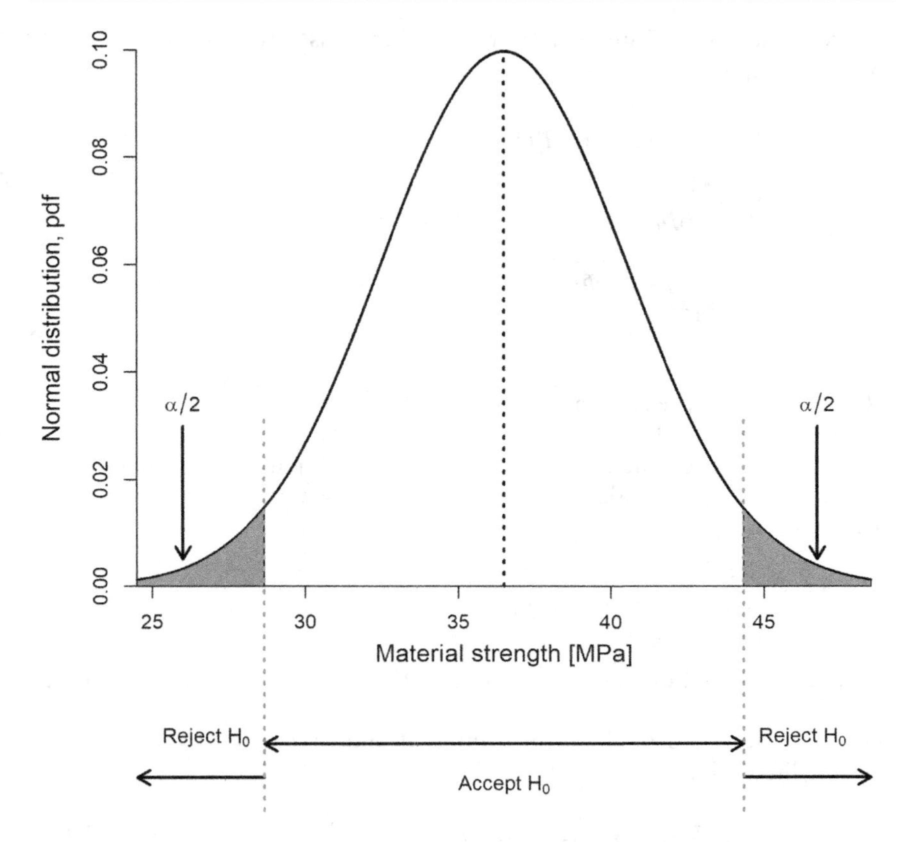

Figure 10.4 Distribution of material strengths under H_0. The grey areas in both end of the distribution represent the critical region where a sample value of $\overline{x}$ results in rejection of H_0.

The question now is what is a reasonable value of c in this case? To answer this, start by considering the probability of a Type I error, which is denoted α and defined as:

$$\alpha = P(\text{Reject}\, H_0 \mid H_0 \text{ is true}) \tag{10.6}$$

By combining Eq. 10.6 with the criterion for rejection of H_0 outlined above, the probability α becomes:

$$P(\overline{x} < \mu_0 - c \cup \overline{x} > \mu_0 + c \mid \mu_0) = P(\overline{x} < \mu_0 - c) + 1 - P(\overline{x} < \mu_0 + c) \tag{10.7}$$

Note that if H_0 is true, then the mean value of the concrete strength is equal to μ_0. Next, as $\overline{x}$ is normally distributed, it is possible to transform each of the terms in Eq. 10.7 into standardised normal distributions using Eq. 2.40, and thus Eq. 10.7 becomes:

$$\alpha = P(\bar{x} < \mu_0 - c) + 1 - P(\bar{x} < \mu_0 + c)$$
$$= \Phi\left(\frac{\mu_0 - c - \mu_0}{\sigma/\sqrt{n}}\right) + 1 - \Phi\left(\frac{\mu_0 + c - \mu_0}{\sigma/\sqrt{n}}\right)$$
$$= \underbrace{\Phi\left(\frac{-c}{\sigma/\sqrt{n}}\right)}_{=\alpha/2} + \underbrace{1 - \Phi\left(\frac{c}{\sigma/\sqrt{n}}\right)}_{=\alpha/2} \tag{10.8}$$

Remember that the standard deviation of $\bar{x}$ is $\sigma / \sqrt{n}$ per the central limit theorem in Eq. 10.4.

Having assigned equal probability $\alpha / 2$ to both parts of the critical interval it is possible to calculate to find an expression of the value for c in Eq. 10.8 as:

$$1 - \Phi\left(\frac{c}{\sigma/\sqrt{n}}\right) = \frac{\alpha}{2} \Rightarrow c = \frac{\sigma}{\sqrt{n}}\Phi^{-1}\left(1 - \frac{\alpha}{2}\right) \tag{10.9}$$

By using the slightly easier notation $\Phi^{-1}\left(1 - \frac{\alpha}{2}\right) = z_{1-\alpha/2}$ it becomes apparent that the width of the critical interval and the significance level α are closely linked as:

$$\text{Reject } H_0 : \bar{x} < \mu_0 - \frac{\sigma}{\sqrt{n}} z_{1-\alpha/2} \cup \bar{x} > \mu_0 + \frac{\sigma}{\sqrt{n}} z_{1-\alpha/2} \tag{10.10}$$

The formulation of the rejection interval in Eq. 10.10 can be reorganised as:

$$\frac{\bar{x} - \mu_0}{\sigma / \sqrt{n}} < -z_{1-\alpha/2} \cup \frac{\bar{x} - \mu_0}{\sigma / \sqrt{n}} > z_{1-\alpha/2} \tag{10.11}$$

The standardised mean value in Eq. 10.11 is most often used as the test statistic which results in the critical interval being defined in terms of the $(1 - \alpha / 2)$ quantile in the standard normal distribution.

Finally, there is no fixed rule for deciding on a value of α, but values of 1%, 5%, and 10% are commonly used. The smaller the value of α, the further the test statistics has to be located from μ_0 for the null hypothesis to be rejected. Thus, a small value of α means only rejecting the null hypothesis if there is very strong evidence for doing so.

10.6 THE STATISTICAL POWER OF A TEST

Choosing a small value of α translates into a small probability of rejecting H_0 if it is true; a Type I error. So why not just pick a very small value of α to prevent a wrong decision? The answer is because a small probability of a Type I error automatically increases the probability of a Type II error; the risk of accepting H_0 even if it is not true. Depending on the problem at hand, the outcome of a Type II error might be more severe than a Type I error. Consider the probability of a Type II error, denoted β, as:

$$\beta = P(\text{Accept}\, H_0 \mid H_0\, \text{is false}) \tag{10.12}$$

This probability is derived by combining Eq. 10.12 with the criterion for accepting H_0 outlined above and the definition of the constant c in Eq. 10.9 while assuming that the true mean value is not μ_0 but some other (not yet specified) value μ. In this case, Eq. 10.12 becomes:

$$\begin{aligned}
\beta &= P\left(\mu_0 - \frac{\sigma}{\sqrt{n}} z_{1-\alpha/2} \le \bar{x} \le \mu_0 + \frac{\sigma}{\sqrt{n}} z_{1-\alpha/2} \mid \mu\right) \\
&= P\left(\bar{x} \le \mu_0 + \frac{\sigma}{\sqrt{n}} z_{1-\alpha/2} \mid \mu\right) - P\left(\bar{x} \le \mu_0 - \frac{\sigma}{\sqrt{n}} z_{1-\alpha/2} \mid \mu\right) \\
&= \Phi\left(\frac{\mu_0 + \frac{\sigma}{\sqrt{n}} z_{1-\alpha/2} - \mu}{\frac{\sigma}{\sqrt{n}}}\right) - \Phi\left(\frac{\mu_0 - \frac{\sigma}{\sqrt{n}} z_{1-\alpha/2} - \mu}{\frac{\sigma}{\sqrt{n}}}\right) \\
&= \Phi\left(z_{1-\alpha/2} + \frac{\mu_0 - \mu}{\frac{\sigma}{\sqrt{n}}}\right) - \Phi\left(-z_{1-\alpha/2} + \frac{\mu_0 - \mu}{\frac{\sigma}{\sqrt{n}}}\right)
\end{aligned} \tag{10.13}$$

Figure 10.5 is a visual definition of β, showing the distribution of $\bar{x}$ under both the null hypothesis (solid lines) and the distribution of $\bar{x}$ under the alternative hypothesis (dashed line). The light grey tail ends of the distribution of the Null represent the critical interval where H_0 will be rejected, even if it is true (Type I error). However, for the case where H_0 is false (the true distribution is the one defined under H_1) the dark grey region represents β, i.e. if $\bar{x}$ falls within this region, we would accept H_0 even if it is not true (Type II error).

Finally, the power of a hypothesis test is defined as the probability of rejecting H_0 if H_0 is indeed not true (a good decision). The power is therefore

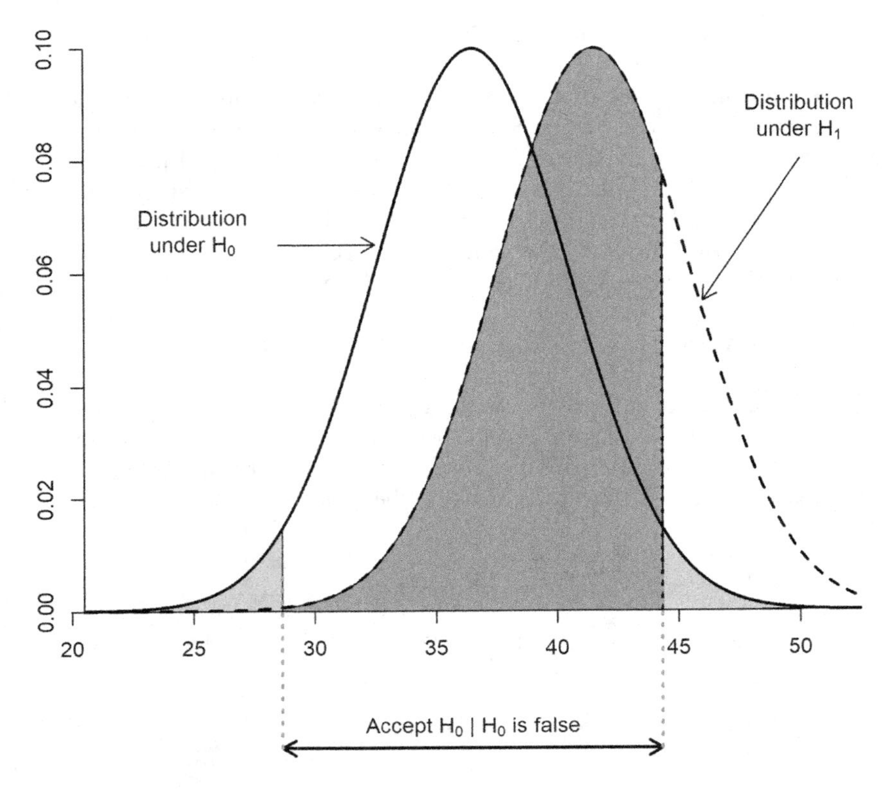

Figure 10.5 Probability β of accepting H_0 if indeed H_0 is false (i.e. $\mu \neq \mu_0$).

defined as 1 minus the probability of a Type II error, so utilising the result in Eq. 10.13 gives the power as:

$$
\begin{aligned}
P(\text{Reject } H_0 \mid H_0 \text{ is false}) &= 1-\beta = 1-P(\text{Accept } H_0 \mid H_0 \text{ is false}) \\
&= 1-\Phi\left(z_{1-\alpha/2}+\frac{\mu_0-\mu}{\frac{\sigma}{\sqrt{n}}}\right)+\Phi\left(-z_{1-\alpha/2}+\frac{\mu_0-\mu}{\frac{\sigma}{\sqrt{n}}}\right) \\
&= \Phi\left(-z_{1-\alpha/2}-\frac{\mu_0-\mu}{\frac{\sigma}{\sqrt{n}}}\right)+\Phi\left(-z_{1-\alpha/2}+\frac{\mu_0-\mu}{\frac{\sigma}{\sqrt{n}}}\right)
\end{aligned}
\tag{10.14}
$$

Figure 10.6 shows the definition of the power (grey shaded area) when considering the two competing distributions under the Null H_0 and the alternative hypothesis H_1.

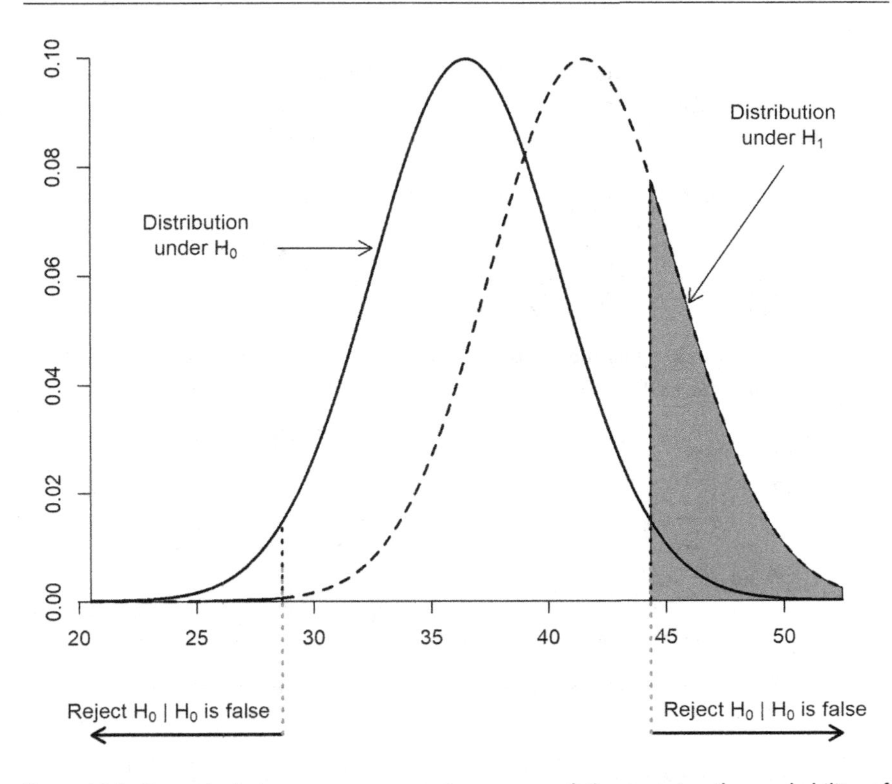

Figure 10.6 Grey shaded areas represent the power of the test, i.e. the probability of rejecting H_0 given H_0 is indeed false (i.e. making the correct decision).

EXAMPLE 10.2 TWO-SIDED HYPOTHESIS TEST

Returning to the example from section 10.1 considering the results of 25 strength tests. According to Figure 10.1 the mean value of the 25 tests is 37.8 MPa. Assuming that the strength is normally distributed with a known standard deviation of the test results $\sigma = 4$, an analyst now wants to investigate if the strength equals the factory specified value of 36.5 MPa at a significance level of $\alpha = 5\%$.

Step one of this analysis is to formulate the null hypothesis and the corresponding alternative hypothesis:

$$H_0 : \mu_0 = 36.5 \text{ MPa}$$

$$H_1 : \mu_0 \neq 36.5 \text{ MPa}$$

The relevant test statistic is the mean value, and the critical interval is defined according to Eq. 10.11 as:

Reject H_0 if:

$$37.8 < 36.5 - \frac{4}{\sqrt{25}} z_{1-0.05/2} \cup 37.8 > 36.5 + \frac{4}{\sqrt{25}} z_{1-0.05/2}$$

$$\frac{37.8 - 36.5}{4 / \sqrt{25}} < -z_{0.975} \cup \frac{37.8 - 36.5}{4 / \sqrt{25}} > z_{0.975}$$

The $(1 - \alpha / 2)$ quantile of the standardised normal distribution can be found using the *NORM.INV* function in EXCEL as shown in the figure below:

Function Arguments ? ×

NORMINV

Probability	0.975	= 0.975
Mean	0	= 0
Standard_dev	1	= 1

= 1.959963985

This function is available for compatibility with Excel 2007 and earlier.
Returns the inverse of the normal cumulative distribution for the specified mean and standard deviation.

Standard_dev is the standard deviation of the distribution, a positive number.

Formula result = 1.959963985

Help on this function OK Cancel

Screenshot of EXCEL function NORM.INV for calculating $z_{0.975}$

From the figure it can be observed that $z_{0.975} = 1.96$

As the value of the test statistics $\frac{37.8 - 36.0}{4/\sqrt{25}} = 2.25 > 1.96$ is larger than the threshold value of 1.96, the test statistic falls within the critical interval, and therefore the null hypothesis should be rejected.

EXAMPLE 10.3 ONE-SIDED HYPOTHESIS TEST

A more meaningful test might consider if the strength of the concrete is below a certain minimum requirement based on the output of the lab tests:

$$H_0 : \mu_0 \leq \mu_0$$

$$H_1 : \mu > \mu_0$$

In this case the test is formulated such that unless there is strong evidence to the contrary, we will accept H_0 (an undesirable outcome). The corresponding critical interval and region of acceptance are given as:

Accept: $\bar{x} < c$

Reject: $\bar{x} \geq c$

The constant c can be determined by considering that we specify a probability α of rejecting H_0 if H_0 is true (Type I error), i.e.:

$$P(\bar{x} > c) = \alpha \rightarrow c = \frac{\sigma}{\sqrt{n}} z_{1-\alpha} + \mu_0$$

This is known as a one-sided test as the critical interval is located exclusively in one (rather than two) sides of the distribution as illustrated in the figure below (note the difference between this figure and Figure 10.4). The critical interval for this one-sided test is given as

$$\bar{x} > \frac{\sigma}{\sqrt{n}} z_{1-\alpha} + \mu_0 \rightarrow \frac{\bar{x} - \mu_0}{\sigma / \sqrt{n}} > z_{1-\alpha}$$

Using the *NORM.INV* function in EXCEL to find $z_{0.95} = 1.64$, the test statistics $\frac{37.8 - 36.0}{4 / \sqrt{25}} = 2.25 < 1.64$ and therefore the null hypothesis should be rejected., i.e. the strength of FAPC exceeds μ_0.

The power of the one-sided test is defined as:

$$1 - \beta = 1 - P(Accept\, H_0 \;|H_0\, is\, false)$$

$$= 1 - P\left(\bar{x} \leq \frac{\sigma}{\sqrt{n}} z_{1-\alpha/2} + \mu_0 \,|\, \mu\right)$$

$$1 - \Phi\left(z_{1-\alpha} + \frac{\mu_0 - \mu}{\sigma / \sqrt{n}}\right)$$

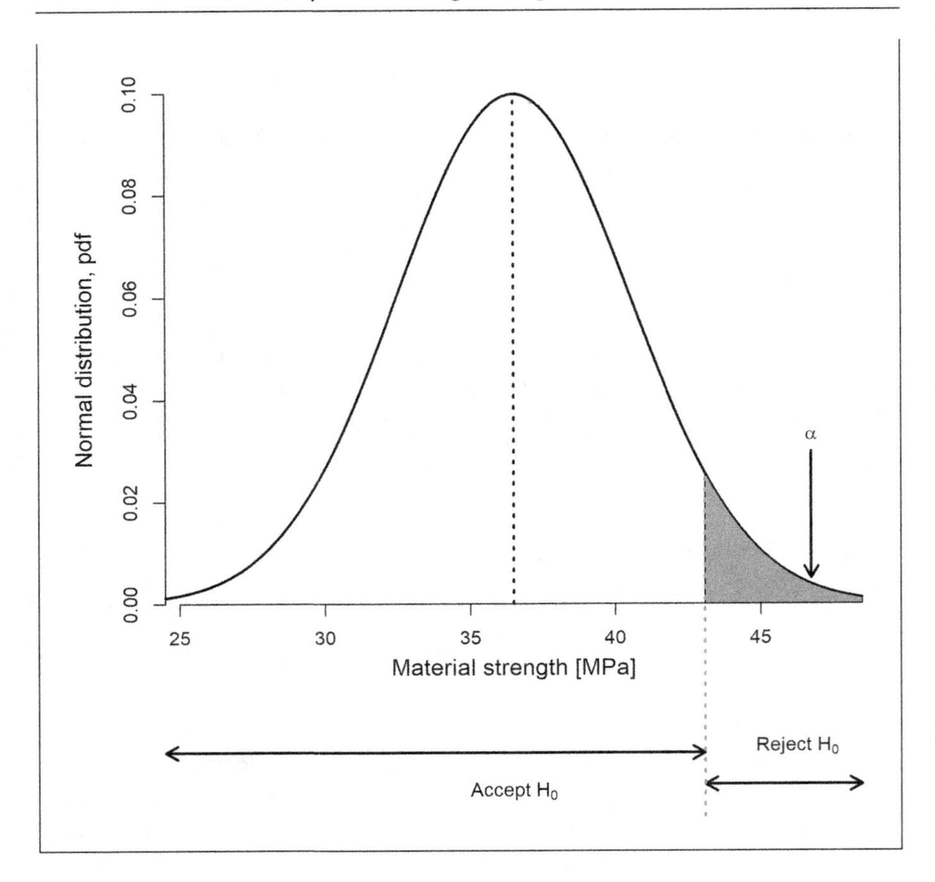

A summary of the different hypotheses and the associated definitions of critical intervals and power are shown in Table 10.2 for the case where the data follows a normal distribution with a known standard deviation but unknown mean value.

The emphasis here is on a single sample. This is because there are also hypothesis tests that allow a comparison between two samples, e.g. comparing the strength of two or more different types of concrete. While these types of tests are often used in practice, they are not discussed further here.

10.7 HYPOTHESIS TESTING WITH UNKNOWN STANDARD DEVIATION

The cases discussed above all assume that the standard deviation of the data is known. When analysing real-world data this is not always the case, and

Table 10.2 Summary of Hypothesis Test for a Single Sample with Unknown Mean Value But Known Standard Deviation

H_0 Hypothesis	*H1: Alternative*	*Test Statistic,* z_0	*Critical Interval*	*Power*
$\mu = \mu_0$	$\mu \neq \mu_0$		$z_0 < z_{\alpha/2}$ OR $z_0 > z_{1-\alpha/2}$	$\Phi\left(-z_{1-\alpha/2} - \frac{\mu_0 - \mu}{\frac{\sigma}{\sqrt{n}}}\right) + \Phi\left(-z_{1-\alpha/2} + \frac{\mu_0 - \mu}{\frac{\sigma}{\sqrt{n}}}\right)$
$\mu \geq \mu_0$	$\mu < \mu_0$	$\frac{\overline{x} - \mu_0}{\sigma / \sqrt{n}}$	$z_0 < z_\alpha$	$\Phi\left(z_\alpha - \frac{\mu - \mu_0}{\sigma / \sqrt{n}}\right)$
$\mu < \mu_0$	$\mu \geq \mu_0$		$z_0 > z_{1-\alpha}$	$1 - \Phi\left(z_{1-\alpha} - \frac{\mu - \mu_0}{\sigma / \sqrt{n}}\right)$

the standard deviation must be estimated from the data. In such cases, the test statistic t for the H_0 hypothesis $\mu = \mu_0$ becomes:

$$t = \frac{\overline{x} - \mu_0}{s / \sqrt{n}} \tag{10.15}$$

Where s is the estimated standard deviation from Eq. 3.1. For this case, the test statistics in Eq. 10.15 no longer follow a normal distribution (as was the case in the examples above when the standard deviation was assumed known) but rather a t-distribution. A more in-depth discussion of the mathematical link between the normal and the t-distribution is beyond the scope of this presentation. Sufficient to say that for large sample sizes the two distributions are very similar, and the normal distribution can be used.

10.8 DEFINITION OF P-VALUES

The final concept to be discussed here is the p-value, which is often described as the probability of obtaining a test result at least as extreme as what was observed assuming the null hypothesis is true. Thus, there is a close link between the p-value and the probability of a Type I error, denoted α and discussed above.

EXAMPLE 10.4 CALCULATING AND INTERPRETING THE *p*-VALUE

In the case of FAPC introduced in section 10.1, an estimate of the mean value of 37.8 MPa was obtained from the 25 tests. The null hypothesis of $\mu_0 = 36$ MPa and $\sigma = 4.0$ MPa was specified. Thus, if the null hypothesis is true, then the probability that the estimated material strength is as least as extreme as 37.8 MPa is:

$$P(\overline{x} > 37.8 | \mu_0 = 36.0) = 1 - P\left(\frac{\overline{x} - 36.0}{4.0 / \sqrt{25}} \leq \frac{37.8 - 36.0}{4.0 / \sqrt{25}}\right)$$
$$= 1 - \Phi(2.25) = 1 - 0.9878 = 0.012$$

Thus, in this case the p-value is 0.012.

Chapter 11

Linear Regression Models

11.1 A SIMPLE LINEAR REGRESSION MODEL

To introduce the concept of simple linear regression and illustrate how it might be useful in civil engineering, consider the following example based on geotechnical data published by Shirur and Hiremath (2014). The saturated California Bearing Ratio (CBR) test is used to ascertain the mechanical strength of a road substrate. A problem with this type of CBR test is that each test takes about a week to complete, and as a result only a limited number of tests can realistically be performed along a proposed road network without causing delays and costs. With only a limited number of tests, there is a risk that the underlying variability of CBR values might not be properly assessed, potentially leading to substandard construction. In an attempt to overcome these problems, a simple mathematical relationship can be established that would enable the prediction of CBR values based on the plasticity index (PI), which can be obtained faster – typically within 30 minutes for an experienced geotechnical engineer. To establish a link between the two soil properties, a total of $n = 20$ soil samples were analysed and values of both CBR and PI were obtained for each sample through a series of tests in a laboratory. The results are shown in Table 11.1. For three of the samples, no values were recorded for the plasticity index, reducing the total sample size from 20 to 17.

As the objective of the data analysis is to predict values of CBR based on values of PI, the CBR is defined as the dependent variable y and PI as the explanatory variable x. Plotting the values of CBR against PI for each of the $n = 17$ soil samples (see Figure 11.1), it can be observed that the 17 points appear to be scattered around a straight line, suggesting that the variation in observed values of CBR can to some extent be explained by the variation in the values of the PI, hence the term *explanatory variable*. For example, a low value of PI appears associated with a low value of CBR and vice versa. There are several reasons why the points in Figure 11.1 do not lie perfectly on a straight line, including experimental errors in determining both CBR and PI measurements, and the existence of geotechnical soil

DOI: 10.1201/9781032700373-11

Table 11.1 Results of Lab Tests for CBR and PI

Sample Number	*PI (%)*	*CBR (%)*	*Sample Number*	*PI (%)*	*CBR (%)*
1	N/A	4.84	11	16	3.94
2	35	1.06	12	18.5	3.28
3	25	2.03	13	24	2.95
4	10	4.18	14	N/A	5.25
5	27	2.79	15	22	4.92
6	20	3.20	16	22.4	3.28
7	24	1.56	17	N/A	4.92
8	28	2.54	18	22	3.12
9	26	2.05	19	31	1.31
10	10.7	3.45	20	20	1.50

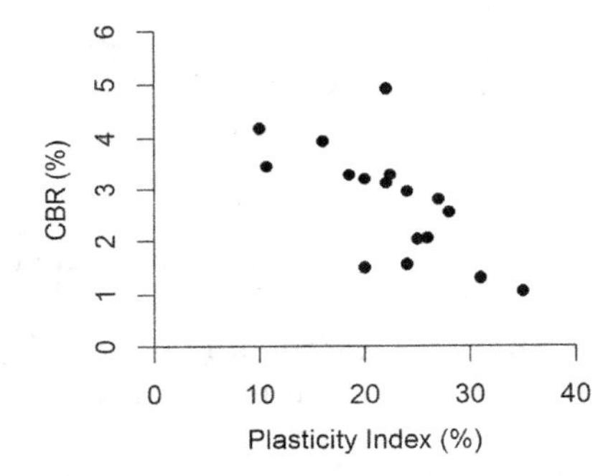

Figure 11.1 Scatter plot showing plasticity index (PI) plotted against values of the California Bearing Ratio (CBR).

properties, other than PI, that determine the CBR values of the soil, but which have not been controlled in this experiment.

Each of the 17 observed values of CBR can now be represented as a contribution from a potential linear relationship with PI and an error term for each data point $(x_i, y_i), i = 1,\ldots n$ measuring the distance from the line to the CBR value. Mathematically, this can be expressed as the basic equation for a straight line plus an error term:

$$y_i = \underbrace{\theta_0 + \theta_1 x_i}_{\text{straight line}} + \underbrace{\varepsilon_i}_{\text{error}}, \quad i = 1,\ldots,n \tag{11.1}$$

where θ_0 and θ_1 are the intercept and slope of a straight line, and ε is a statistical *error* (or residual), which is assumed to be a normally distributed random variable with mean zero and variance σ^2. The first two terms on the right-hand side of Eq. 11.1 represent the as of yet undefined, best straight line through the data, and the errors represent the vertical distance between the line and each observation. Thus, for each set of observations (x_i, y_i), $i = 1, \ldots n$ the observed CBR value is a function of the explanatory variable PI and an error component, ε_i representing how much the observation deviates from the straight line, i.e.:

$$\varepsilon_i = y_i - (\theta_0 + \theta_1 x_i), \qquad i = 1, \cdots, n \tag{11.2}$$

The ε_i term defined in Eq. 11.2 is known as a *residual*, calculated as the difference between the observation and the regression line. These residuals are the key to estimating the intercept and slope (θ_0, θ_1) of the straight line that provides the best line through the data points. Using a technique known as the least squares method, the values of (θ_0, θ_1) are sought which minimise the squared sum of the n residuals defined in Eq. 11.2, here expressed through the function $S(\theta_0, \theta_1)$ as:

$$S(\theta_0, \theta_1) = \sum_{i=1}^{n} \varepsilon_i^2 = \sum_{i=1}^{n} (y_i - \theta_0 - \theta_1 x_i)^2 \tag{11.3}$$

The function S in Eq. 11.3 is minimised by differentiating with respect to both θ_0 and θ_1 and setting the resulting derivatives equal to zero as:

$$\begin{aligned} \frac{\partial S}{\partial \theta_0} &= -2\sum_{i=1}^{n} (y_i - \theta_0 - \theta_1 x_i) = 0 \\ \frac{\partial S}{\partial \theta_1} &= -2\sum_{i=1}^{n} x_i (y_i - \theta_0 - \theta_1 x_i) = 0 \end{aligned} \tag{11.4}$$

The solution to the two Eqs. 11.4 results in the following expressions for estimating the intercept and the slope of the regression line:

$$\begin{aligned} \hat{\theta}_0 &= \frac{\left(\sum_{i=1}^{n} x_i^2\right)\left(\sum_{i=1}^{n} y_i\right) - \left(\sum_{i=1}^{n} x_i\right)\left(\sum_{i=1}^{n} x_i y_i\right)}{n\sum_{i=1}^{n} x_i^2 - \left(\sum_{i=1}^{n} x_i\right)^2} \\ \hat{\theta}_1 &= \frac{n\left(\sum_{i=1}^{n} x_i y_i\right) - \left(\sum_{i=1}^{n} x_i\right)\left(\sum_{i=1}^{n} y_i\right)}{n\sum_{i=1}^{n} x_i^2 - \left(\sum_{i=1}^{n} x_i\right)^2} \end{aligned} \tag{11.5}$$

The "$\hat{\ }$" notation is used to indicate that the two expressions in Eq. 11.5 are estimators, which can be thought of as a rule for calculating an estimate of a given quantity based on observed data.

Returning to the example with the $n = 17$ soil samples, the line through the points that gives the smallest squared difference between the line and the points can now be calculated using the formulas given in Eq. 11.5. In

Example 11.1 the slope and intercept (θ_0 and θ_1) of the best fit line through the $n = 17$ soil observations are derived.

If a new value of the explanatory variable x becomes available (e.g. the plasticity index is determined for a new soil sample where CBR has not been measured), then a prediction of the dependent variable y (the required CBR value) can be made as:

$$\hat{y} = \hat{\theta}_0 + \hat{\theta}_1 x \tag{11.6}$$

where, again, the hat signifies an estimator.

EXAMPLE 11.1 ESTIMATING REGRESSION MODEL PARAMETERS

From the data in Table 11.1 the following sums are calculated for use in the parameter estimators in Eqs. 11.5:

$$\sum_{i=1}^{17} x_i = 381.6$$

$$\sum_{i=1}^{17} x_i^2 = 9234.5$$

$$\sum_{i=1}^{17} y_i = 47.16$$

$$\sum_{i=1}^{17} y_i \; x_i = 983.24$$

which gives the following two regression parameters:

$$\theta_0 = \frac{9234.5 \cdot 47.16 - 381.6 \cdot 983.24}{17 \cdot 9234.5 - (381.6)^2} = 5.304$$

$$\theta_1 = \frac{17 \cdot 983.24 - 381.6 \cdot 47.46}{17 \cdot 9234.5 - (381.6)^2} = -0.113$$

Therefore, the regression line that will enable CBR to predict from values of PI is given as:

$$CBR = -0.113 \times PI + 5.304$$

and is shown in the figure below plotted together with the 17 data points.

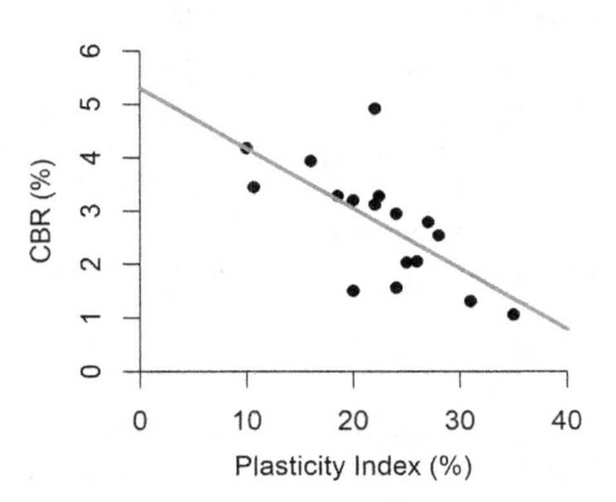

The result is a straight line that fits through the cloud of points, representing the optimal least squares solution.

Notice in Example 11.1 how the regression line (dark grey line) does not give an exact fit to any one observation but rather represents the line that results in the lowest values of the summed squared differences between the line and the data points (hence the name of the fitting technique, least squares). This leads to an important assumption in linear regression models, that the residuals defined in Eq. 11.2 are assumed to be independent of each other and follow a normal distribution with zero mean value and a standard deviation σ_ε representing the degree of scatter of the observations around the best fitting line (as defined by $\hat{\theta}_0$ and $\hat{\theta}_1$). The standard deviation (or variance) is estimated as the sum of squared residuals divided by the number of observations minus 2 degrees of freedom (because the two regression parameters have already been estimated), i.e.:

$$\hat{\sigma}_\varepsilon^2 = \frac{1}{n-2}\sum_{i=1}^{n}\left(y_i - \hat{\theta}_0 - \hat{\theta}_1 x_i\right)^2 \tag{11.7}$$

The residual variance plays an important role when assessing (1) the strength of evidence of a statistical relationship between x and y (also known as statistical significance, see Chapter 10), and (2) the uncertainty of a prediction of y based on a future value of x (see section 11.4).

11.2 INVESTIGATING THE FIT OF A REGRESSION MODEL

An assessment of how well the estimated regression model fits the data is also known as investigating the goodness-of-fit of the regression model. The goodness-of-fit assessment can be made using both graphical and numerical methods.

11.2.1 Visual Assessment of Goodness-of-Fit

Model assessment is conducted after a regression model is fitted and significance checks have been conducted (see section 11.3). It focusses primarily on assessing if the residuals are independent, have equal variance (homoscedasticity), and are normally distributed. As a reminder, the residuals ε_i are defined as the difference between the observed y_i and the corresponding value $\hat{y}_i$ obtained from the regression model using Eq. 11.6, i.e.:

$$\varepsilon_i = y_i - \hat{y}_i = y_i - \left(\hat{\theta}_0 + \hat{\theta}_1 x_i\right) \tag{11.8}$$

If the residuals are plotted against the exploratory variable x_i, then the residuals should ideally show no relation to the x values, and the plot should look like a cloud of random points. Next, the distributional assumption of the residuals is checked. The assumption that residuals are normally distributed with mean zero and variance σ_ε^2 can be checked using normal probability paper (see Chapter 4), where the ranked residuals are plotted against their expected value expressed as a quantile.

EXAMPLE 11.2 RESIDUAL PLOT FOR INDIAN SOILS DATA

For each of the $n = 17$ data points used to derive the regression model relating CBR to PI for the Indian soils data, the corresponding residual ε_i is calculated as

$$\varepsilon_i = CBR_i - \left(-0.113 \cdot PI_i + 5.304\right), i = 1, \ldots 17$$

and plotted against the corresponding values of the plasticity index (PI) as shown in the figure.

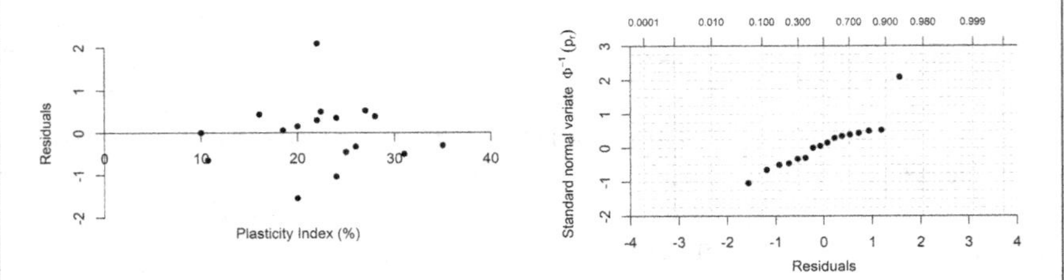

Figure: Residuals plotted against PI (%) (left), and normal probability plot of residuals (right).

The plot shows that the residuals appear to be reasonably randomly scattered around zero with no obvious systematic relationship between the residuals and the values of the plasticity index, which is reassuring. There is one residual which has a relatively large value. In such cases it might be useful, if possible, to re-examine the experiment to check if the data can be considered error-free.

The normal probability plot on the right-hand side shows that most of the residuals fall approximately on a straight line in the frequency plot, apart from the same outlier as discussed above. This confirms that further investigation of this outlier should be conducted before a final model is developed. If, and only if, after closer inspection the data point is considered suspect then it should be omitted from the analysis. The problem is that once removed, the second largest deviation will now become the largest deviation and there might be a temptation to remove this data point as well. It is important to be careful not to remove data points to reduce the variation if this means removing valuable information and thus masking the true variability in the data.

11.2.2 Numerical Assessment of Goodness-of-Fit

A numerical evaluation of how well the regression line fits the observed data can be based on considerations of how much of the total variability in the original data y_i can be explained by the values of the explanatory variable x_i via the regression model. First, the total sum of squares is calculated by considering how much the data are scattered around the mean value, i.e.:

$$S_{yy} = \sum_{i-1}^{n} (y_i - \bar{y})^2 \tag{11.9}$$

where $\bar{y}$ is the mean value of the n observations defined as:

$$\bar{y} = \frac{1}{n}\sum_{i=1}^{n} y_i \tag{11.10}$$

Next, consider the observation y_i for which the regression model would provide an estimated value of $\hat{y}_i = \hat{\theta}_0 + \hat{\theta}_1 x_i$. With reference to Figure 11.2 which illustrates the differences for a single point, the value of the observation y_i can be expressed in terms of both $\hat{y}_i$ and the mean value $\bar{y}$ as:

$$\begin{aligned} y_i &= \bar{y} + (y_i - \hat{y}_i) + (\hat{y}_i - \bar{y}) \Leftrightarrow \\ y_i - \bar{y} &= (y_i - \hat{y}_i) + (\hat{y}_i - \bar{y}) \end{aligned} \tag{11.11}$$

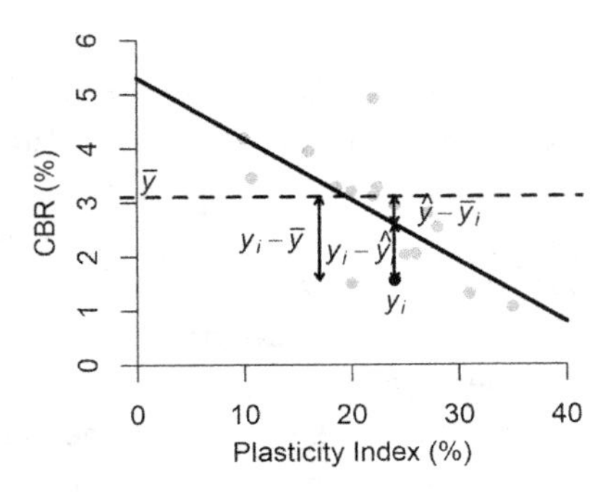

Figure 11.2 Geometrical definition of the differences used in Eq. 11.11 to define variability. The horizonal dashed line represents the mean value $\overline{y}$.

The total variance S_{yy} can now be reformulated by first squaring Eq. 11.11 and then summing over all n data points, the following expression is obtained:

$$\sum_{i=1}^{n}(y_i-\overline{y})^2=\sum_{i=1}^{n}(y_i-\hat{y}_i)^2+\sum_{i=1}^{n}(\hat{y}_i-\overline{y})^2+2\sum_{i=1}^{n}(y_i-\hat{y}_i)(\hat{y}_i-\overline{y}) \tag{11.12}$$

where the last term can be shown to be equal to zero (see the appendix), and the left-hand side is equal to Eq. 11.9, i.e.:

$$S_{yy}=\sum_{i=1}^{n}\ (y_i-\hat{y}_i)^2+\sum_{i=1}^{n}\ (\hat{y}_i-\overline{y})^2 \tag{11.13}$$

This means that the total variation in the data (Eq. 11.13, left-hand side) can be explained partly by the regression model (second term on the right-hand side) plus the noise that is left over, represented by the sum of the squared residuals (first term on the right-hand side). An often-used measure of the quality of the fit of the regression line is the *coefficient of determination*, R^2, defined as the ratio between the variation explained by the regression model and the total variation of the observations, i.e.:

$$R^2=\frac{\sum_{i=1}^{n}(\hat{y}_i-\overline{y})^2}{\sum_{i=1}^{n}(y_i-\overline{y})^2} \tag{11.14}$$

By substituting Eq. 11.13 into the expression above for R^2, the more frequently used version is given as

$$R^2=1-\frac{\sum_{i=1}^{n}(y_i-\hat{y})^2}{\sum_{i=1}^{n}(y_i-\overline{y})^2} \tag{11.15}$$

The better the fit of the regression line is, the smaller the differences between the observations and the line are, i.e. lower values of the sum $\sum_{i=1}^{n}(y_i - \hat{y})^2$. If all the observations are located exactly on the line of best fit (i.e. a perfect fit), then that would give $R^2 = 1$ while an increasingly large scatter of the data around the line would result in increasingly values of R^2 less than 1.

EXAMPLE 11.3 LINEAR REGRESSION USING EXCEL

Developing a regression model and calculating the associated R^2 *coefficient of determination* is straightforward using a spreadsheet model like EXCEL. It is common to organise the data in columns such as the PI and CBR data from Table 11.1, as shown in Figure A.

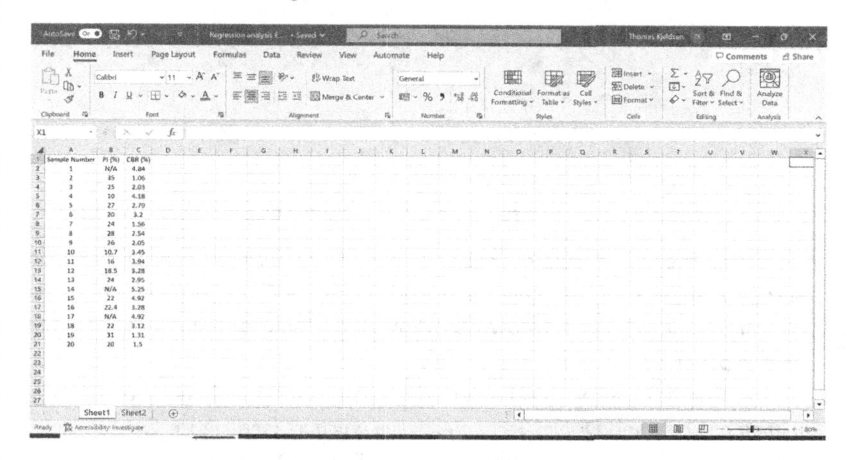

Sample Number	PI (%)	CBR (%)
1	N/A	4.84
2	35	1.06
3	25	2.03
4	10	4.18
5	27	2.79
6	20	3.2
7	24	1.56
8	28	2.54
9	26	2.05
10	10.7	3.45
11	16	3.94
12	18.5	3.28
13	24	2.95
14	N/A	5.25
15	22	4.92
16	22.4	3.28
17	N/A	4.92
18	22	3.12
19	31	1.31
20	20	1.5

Figure A Geotechnical data (PI and CBR) from Table 11.1 organised in columns in EXCEL.

It is possible to estimate the two model parameters manually by evaluating the individual sums as shown in Example 11.1 by adding additional columns to the spreadsheet. However, the task is made easier by using the EXCEL's built-in functions.

First, remove the three rows containing N/A data points, as these are cannot be used in the numerical analysis, i.e. remove the rows containing sample 1, 14, and 17 leaving only 17 data points of the original 20.

Next, highlight the two in the two columns named PI and CBR as shown in the Figure A above.

Move the cursor to the *Insert* menu and select the *Insert scatter* (X, Y) option (see circled options in Figure B, left).

Selecting the *Insert Scatter* (X, Y) option will provide a set of plotting options as shown in Figure B (right). Chose the *Scatter* option (circled on Figure B). Other options are available which will draw lines between the individual points, which is not the objective of the regression analysis.

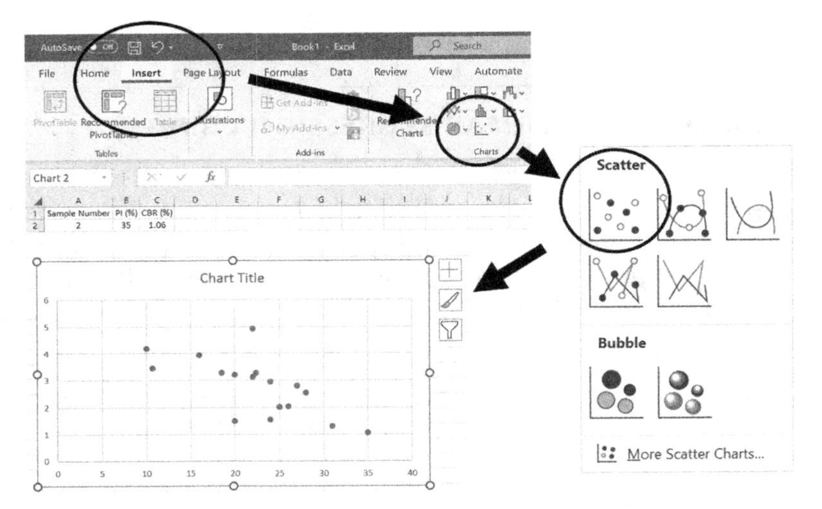

Figure B Follow the black arrows to Insert a $X - Y$ scatter diagram on the worksheet.

Inserting the $X - Y$ scatter will also initiate the *Format Trendline* panel on the right-hand side of the screen as shown in Figure C. Note that it might be necessary to scroll up and down to view all available options in the *Format Trendline* panel.

In the *Format Trendline* panel, select the *Linear* option and check the *Display Equation on chart* and *Display R-squared value on chart* (see Figure C).

The end-result of the process is shown in Figure D, where the least squares regression line has been added to the initial $X - Y$ scatter plot as well as the estimated model parameters and the R^2 coefficient of determination. Cross-checking the result of the automated EXCEL procedure with the manual calculations in Example 11.1 confirms that EXCEL has produced the correct results. Of course, it is necessary to add appropriate axis labels before disseminating the results to a wider audience (see Chapter 4 on the design of graphs).

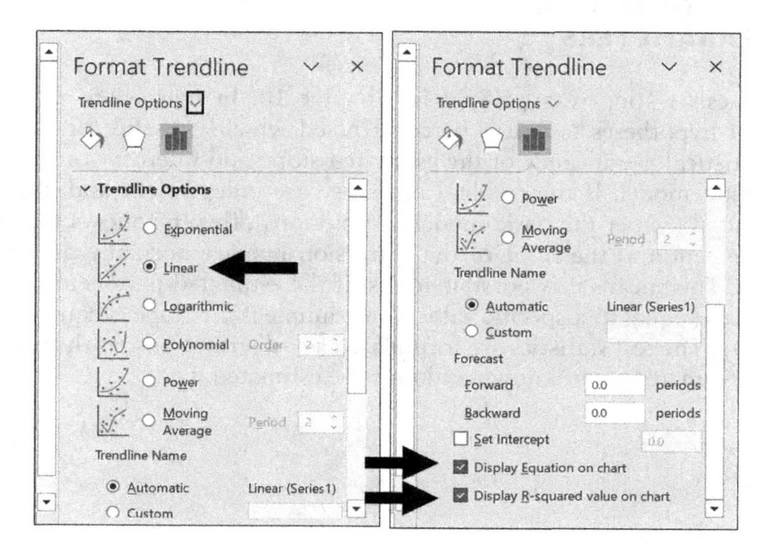

Figure C Select the Linear option and tick the Display Equation on chart and Display R-squared value on chart.

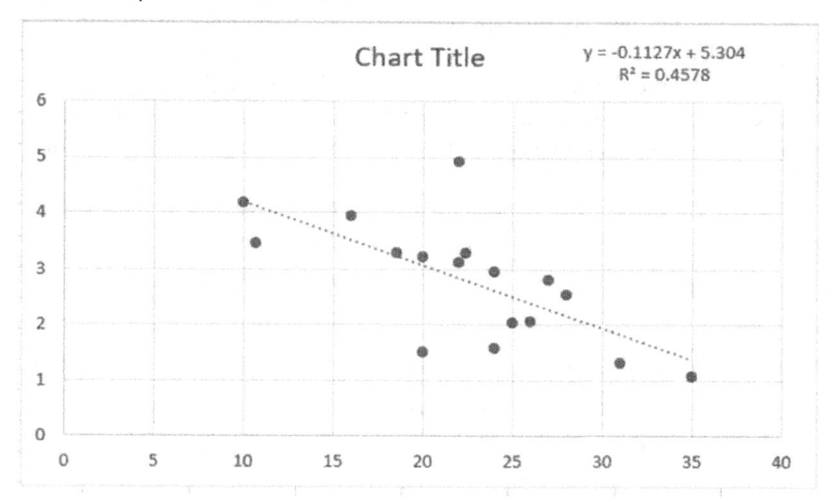

Figure D Linear regression model added to $X-Y$ scatter, including the equation and the R^2.

11.3 TESTING SIGNIFICANCE OF REGRESSION PARAMETERS

Hypothesis testing was covered in Chapter 10. In regression modelling, a set of hypotheses tests can be constructed which is useful for assessing the statistical significance of the estimated slope and intercept of the fitted regression model. If the residual errors ε_i are independent and normally distributed (as per the basic model assumptions), then it follows that both the intercept and the slope of the regression line are normally distributed as well. This means it is possible to test if the estimated parameters can be considered equal to a specific value, for example if the slope is equal to zero $(\theta_0 = 0)$. The test statistics are formulated as t-distributions, as the variance of the estimates are unknown and must be estimated, i.e.

$$\frac{\hat{\theta}_0 - \theta_0}{\sqrt{V(\hat{\theta}_0)}} \sim t_{N-2}$$
$$\frac{\hat{\theta}_1 - \theta_1}{\sqrt{V(\hat{\theta}_1)}} \sim t_{N-2} \tag{11.16}$$

where t_{n-2} is Student's t-distribution with $n-2$ degrees of freedom (n is the number of observations used in the regression analysis) and $V(\hat{\theta}_0)$ and $V(\hat{\theta}_1)$ are obtained directly from Eq. 11.17 below with the residual error variance estimated from Eq. 11.7. Evaluation of the test statistics in Eq. 11.16 requires the variance of the two estimated model parameters, defined as:

$$V(\hat{\theta}_o) = \frac{\sigma_\varepsilon^2 \sum_{i=1}^N x_i^2}{N\sum_{i=1}^N x_i^2 - \left(\sum_{i=1}^N x_i\right)^2}$$
$$V(\hat{\theta}_1) = \frac{N\sigma_\varepsilon^2}{N\sum_{i=1}^N x_i^2 - \left(N\sum_{i=1}^N x_i\right)^2} \tag{11.17}$$
$$Cov(\hat{\theta}_0, \hat{\theta}_1) = \frac{-\sigma_\varepsilon^2 \sum_{i=1}^N x_i}{N\sum_{i=1}^N x_i^2 - \left(\sum_{i=1}^N x_i\right)^2}$$

All three variance and covariance expressions above depend on only the explanatory variables x_i which are known, and the variance of the residuals σ_ε^2 which is unknown and therefore must be estimated.

A common task is to investigate if the slope θ_1 can be assumed equal to zero. If indeed the null hypothesis $\theta_1 = 0$ is accepted, then the regression model in Eq. 11.1 effectively reduces to $y_i = \theta_0 + \varepsilon_i$ and there is no evidence for a significant relationship between x and y.

EXAMPLE 11.4 HYPOTHESIS TESTING

It is common to conduct a hypothesis test to ascertain if the estimated model parameters $\hat{\theta}_0$ and $\hat{\theta}_1$ can be assumed statistically different from zero, i.e. a null hypothesis $\theta_0 = 0$ and/or $\theta_1 = 0$. Most statistical software packages summarise this set of hypothesis tests in a table format, such as the following, containing the results based on the $n = 17$ data points.

Summary table for significance test of model parameters

Parameter	*Estimate*	*Std. Dev.*	*t-Value*	*p-Value*
θ_0	5.304	0.7381	7.186	3.14×10^{-6}
θ_1	−0.113	0.0317	−3.559	0.0029

First, the model parameters $\hat{\theta}_0$ and $\hat{\theta}_1$ and their associated standard deviation ($\sqrt{\text{variance}}$) are estimated as reported in Example 11.1. Next, the test statistics (*t*-values) are calculated using Eq. 11.16 by assuming that the true values of the model parameters are $\theta_0 = 0$ and $\theta_1 = 0$, reflecting the established null hypothesis. Finally, the *p*-value is derived, representing the probability of observing a test statistic more extreme than the observed value if the null hypothesis is true (see Example 10.4). It is common to use a two-sided *t*-test, considering the possibility of both positive and negative deviations from the null hypothesis. The results in the table show that the p-values for both the intercept and slope are below 0.05, which means both parameters are statistically significant and should be included in the regression model.

11.4 FUTURE PREDICTIONS USING LINEAR REGRESSION

Once a linear regression model has been developed based on statistical analysis of available data, the model can be used for making prediction of values of y based on future values of the explanatory variable x. For example, the model-developed Example 11.1 can now be used to predict values of CBR based on values of PI without necessarily going through the trouble of doing the CBR test. In a general case, assume that a future (subscript f) value of the explanatory variable is denoted x_f, then the corresponding predicted (or estimated) value will be denoted y_f and given as:

$$\hat{y}_f = \hat{\theta}_0 + \hat{\theta}_1 x_f \tag{11.18}$$

Even if the actual value of the value y_f is unknown (remember $\hat{y}_f$ is only an estimate), it is assumed that it follows the same basic statistical model outlined in Eq. 11.1 and therefore can be defined as a combination of the linear component and the residual as:

$$y_f = \theta_0 + \theta_1 x_f + \varepsilon_f \tag{11.19}$$

Notice here that the expression for $\hat{y}_f$ in Eq. 11.18 is a prediction of y_f, whereas the expression in Eq. 11.19 is the actual (yet unknown) value of the y_f; an important distinction. The difference between the true value and our best guess is therefore obtained by subtracting the two expressions above as:

$$\begin{aligned} y_f - \hat{y}_f &= \theta_0 + \theta_1 x_f + \varepsilon_f - \left(\hat{\theta}_0 + \hat{\theta}_1 x_f\right) \\ &= \left(\theta_0 - \hat{\theta}_1\right) + \left(\theta_0 - \hat{\theta}_1\right) x_f + \varepsilon_f \end{aligned} \tag{11.20}$$

The difference in Eq. 11.20 is a combination of errors introduced into the prediction because (1) the prediction has to be made with the estimated parameters $\hat{\theta}_0$ and $\hat{\theta}_1$ rather than the true (but unknown) values θ_0 and θ_1, and (2) the unavoidable "error" ε_f because factors influencing the values of y beyond what is measured using the explanatory variable x have not been considered in this simple model.

The variance of the prediction is defined as the variance of the difference $y_f - \hat{y}_f$, i.e.

$$\begin{aligned} V(y_f - \hat{y}_f) &= V\left((\theta_0 - \hat{\theta}_0) + (\theta_1 - \hat{\theta}_1) x_f + \varepsilon_f\right) \\ &= V(\hat{\theta}_0) + V(\hat{\theta}_1)(x_f)^2 + 2x_f Cov(\hat{\theta}_0, \hat{\theta}_1) + \sigma^2 \end{aligned} \tag{11.21}$$

Where independence between the residual errors and the model parameter estimates have been assumed. Inserting the expressions for $V\left(\hat{\theta}_0\right)$, $V\left(\hat{\theta}_1\right)$, and $Cov\left(\hat{\theta}_0, \hat{\theta}_1\right)$ from Eq. 11.17 the variance of the prediction error can be defined as:

$$\begin{aligned} V\left(y_f - \hat{y}_f\right) &= \sigma^2 \left(1 + \frac{1}{n} + \frac{\left(x_f - \overline{x}\right)^2}{\sum_{i=1}^{N} \left(x_i - \overline{x}\right)^2}\right) \\ &= \sigma^2 \left(1 + \frac{1}{n} + \frac{\left(x_f - \overline{x}\right)^2}{Sxx}\right) \end{aligned} \tag{11.22}$$

Where the denominator Sxx is defined as $\sum_{i=1}^{n} \left(x_i - \overline{x}\right)^2$. If the number of observations is large and the difference between x_f and $\overline{x}$ is small, then the prediction variance might be conveniently approximated as simply the variance of the residuals, i.e.:

$$V\left(y_f - \hat{y}_f\right) \approx \sigma_\varepsilon^2 \tag{11.23}$$

The notation $V(y_f - \hat{y}_f)$ can be reduced to $V(\hat{y}_f)$ as y_f is essentially a constant value.

EXAMPLE 11.5 PREDICTING A FUTURE VALUE OF CBR, INCLUDING UNCERTAINTY

Starting from the regression model developed in Example 11.1 relating CNR to PI and consider a future value of $PI_f = 12\%$, which gives a predicted value of $\widehat{CBR}_f$ as:

$$\widehat{CBR}_f = -0.113 \times PI_f + 5.304 = -0.113 \times 12 + 5.304 = 3.952$$

Here the hat notation on *CBR* is used to indicate that this value is estimated from the regression model. To quantify the uncertainty of this prediction, it is necessary to first estimate the variance of the residuals using Eq. 11.7 to the current example data:

$$\sigma_\varepsilon^2 = \frac{1}{17-2}\sum_{i=1}^{17}\left(CBI_i - (-0.113 \times PI_i + 5.304)\right)^2 = 0.6706$$

Next, the variance of the prediction of CBI_f is calculated using Eq. 11.22, noticing that the average value of the explanatory variable PI obtained from the 17 observations is $\overline{PI} = 22.45\%$ and *Sxx* calculated as:

$$Sxx = \sum_{i=1}^{17}\left(PI_i - \overline{PI}\right)^2 = 668.70$$

It is now possible to calculate the variance of the prediction as:

$$V\left(\widehat{CBI}_f\right) = 0.6706\left(1 + \frac{1}{17} + \frac{(12-22.45)^2}{668.70}\right) = 0.8195$$

Finally, the uncertainty of the prediction can be reported as a 95% confidence interval by assuming the predicted value of CBI to be normally distributed with a mean value equal to the prediction and a standard deviation equal to the square root of the variance calculated above. The 95% confidence interval spans approximately 2 times the standard deviation on both sides of the mean value, i.e.:

$$\left[\widehat{CBI}_f - 2\sqrt{V\left(\widehat{CBI}_f\right)}\ ;\ \widehat{CBI}_f + 2\sqrt{V\left(\widehat{CBI}_f\right)}\right]$$

Inserting the results from above gives a final 95% confidence interval for the prediction of CBI as:

$$[2.14 : 5.76]$$

The prediction and associated confidence interval at PI=12% are shown in the figure below created in EXCEL.

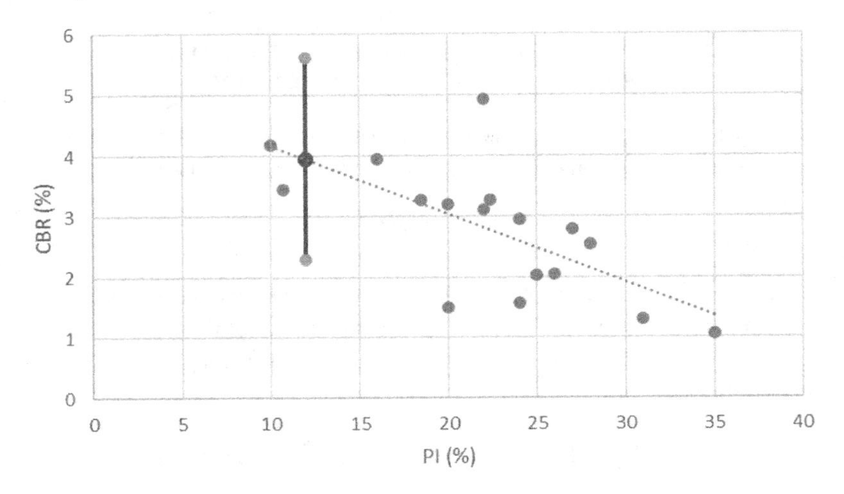

The width of the confidence interval is determined by the size of the variance estimate, which itself is a function of record length n and the residual variance σ_{ε}^{2}. If the 95% confidence interval is deemed to be too wide, then the options are (1) add more data to the analysis (increase n), (2) develop a new regression model that fits the data better (reduce the residual variance), or (3) do both (1) and (2).

REFERENCE

Shirur, N. B. and Hiremath, S. G. (2014). Establishing relationship between CBR value and physical properties of soil. *IOSR Journal of Mechanical and Civil Engineering*, 11(5), pp. 26–30.

Appendix

The derivation of Eq. 11.13 was based on the following assumption:

$$2\sum_{i=1}^{n}(y_i - \hat{y}_i)(\hat{y}_i - \bar{y}) = 0 \tag{11.24}$$

To show that this assumption is valid, consider that the predicted value is defined in Eq. 11.6 as $\hat{y}_i = \hat{\theta}_0 + \hat{\theta}_1 x_i$, and the definition of a residual from Eq. 11.2 as $\varepsilon_i = y_i - \hat{y}_i$. ignoring the "2" in front of the summation sign, and Eq. 11.24 can be reformulated as:

$$\sum_{i=1}^{n}\varepsilon_i\left(\hat{\theta}_0 + \hat{\theta}_1 x_i - \bar{y}\right) = 0 \tag{11.25}$$

As $\hat{\theta}_0$, $\hat{\theta}_1$ and $\bar{y}$ are all constants, Eq. 11.25 can be reorganised to give:

$$\left(\hat{\theta}_0 - \bar{y}\right)\sum_{i=1}^{n}\varepsilon_i + \hat{\theta}_1\sum_{i=1}^{n}\varepsilon_i x_i = 0 \tag{11.26}$$

The next step is to realise that from the derivation of the optimal values of $\hat{\theta}_0$ and $\hat{\theta}_1$ using the least squares method in Eq. 11.5, the following relationships exist:

$$\begin{aligned}\sum_{i=1}^{n}\varepsilon_i &= \sum_{i=1}^{n}\left(y_i - \hat{\theta}_0 - \hat{\theta}_1 x_i\right) = 0\\ \sum_{i=1}^{n}\varepsilon_i x_i &= \sum_{i=1}^{n}\left(y_i - \hat{\theta}_0 - \hat{\theta}_1 x_i\right)x_i = 0\end{aligned} \tag{11.27}$$

When combined with Eq. 11.26, it can be observed that the expression in Eq. 11.24 is therefore always true.

Chapter 12

Low Probability, High Impact Events

12.1 A TYPOLOGY OF LOW PROBABILITY EVENTS

In civil engineering, as in many other aspects of life, we are often concerned about events that have a seemingly low probability of occurring but would have catastrophic consequences should they materialise. Examples include total collapse of a bridge or building, nuclear disaster, economic meltdown, deadly pandemics – the list is seemingly endless. But it could also be a positive event such as winning the lottery. As low probability events are by their very definition rare, there are often little or no prior data or empirical evidence on which to build statistical models such as those covered in the previous chapters. For example, with only limited or no data it is not possible to build credible models using concepts such as mean, standard deviation, or the choice of distribution type. Making quantitative probability statements about these types of very impactful events is therefore challenging. Communication to a general audience about the probability and impact of such events is also challenging, as shall be discussed in Chapter 13.

This chapter will introduce different ideas on how to classify and manage low probability, high impact events. In particular, the following sections will discuss ideas of surprising events, record-breaker and black swan events, before reviewing a general framework for risk management.

12.2 LOW PROBABILITY EVENTS: SHOULD WE BE SURPRISED?

Recall the discussion of sample space and probability models in Chapter 2. A low probability event is simply an event A located within the sample space Ω for which the associated probability $p(A)$ is low compared to other events within the same sample space. A low probability is not in itself a reason to be surprised by an event and does not make it an interesting event, as discussed by Weaver (1948). Consider for example a game of cards where

DOI: 10.1201/9781032700373-12

each player is dealt 13 cards from a deck of 52 cards. As the order in which the cards are dealt is immaterial, there are $\binom{52}{13} = 650{,}013{,}559{,}600$ different combinations; that is more than 650 billion different combinations of cards, each one being equally likely. As each combination is equally likely, there is no scientific justification for being surprised by a hand consisting of all 13 spades more than any other combination; it is rare and improbable but not a surprising event. Contrast this example with the case of flipping a coin where the edge is thick enough to allow a one in a billion chance that it balances on the edge. The corresponding probability of either side of the disk is half minus half of one in a billion. We would expect that in most cases the disk would land on one of the two sides. But in the very rare instance where it balances on the edge is both rare, improbable, surprising, and interesting, even if it is about 650 times more likely than getting a hand full of spades. What makes the event surprising is not necessarily the low probability, but rather the small probability in contrast to other more likely events. Whether or not an event is considered interesting depends on the perspective of the observed; what is interesting to you might be considered not interesting at all by others.

12.3 RECORD-BREAKING EVENTS

The term *record-breaking* is often evoked in connection with meteorological or economic phenomena. For example, a record-breaking heat wave or rainfall, or record-breaking gains or losses on the stock market.

To illustrate the idea of record-breakers, Figure 12.1 shows the annual maximum series of peak flow (m^3/s) recorded by the Environment Agency

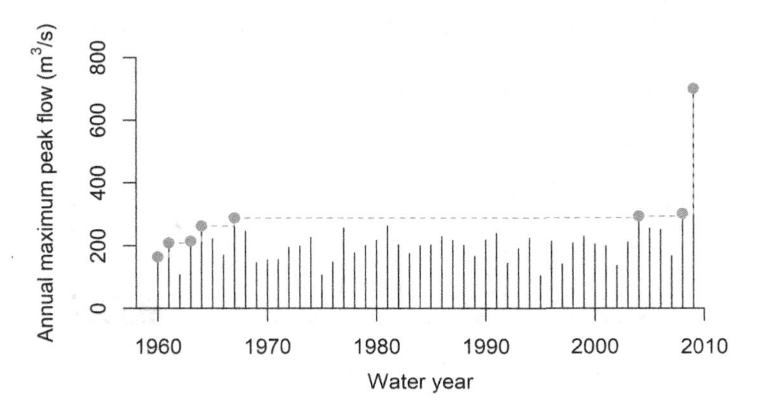

Figure 12.1 Annual maximum peak flow at gauging station 75002. Red points denote record-breakers.

on the River Derwent at Camerton in the English Lake District, Cumbria (gauging station number 75002), where eight record-breakers have been observed over a 50-year period. This series consists of 60 years of data; the record-breaking events are highlighted and can be seen to cluster at the beginning and the end of the record.

Generally, in a series of n observations a record-breaker is defined an event that exceeds all previous observations. Consider a time series of n events $x_1, x_2, \ldots, x_n$, (where x_1 is the first event and x_n is the most recent event); then the technical definition of a record-breaker is that the most recent observation exceeds all previous events, i.e.:

$$x_n > max(x_1, x_2, \ldots, x_{n-1}) \tag{12.1}$$

it is also possible to define record-breakers in terms of being a record-breaking low, i.e.:

$$x_n < min(x_1, x_2, \ldots, x_{n-1}) \tag{12.2}$$

In any record the first observation is always counted as a record-breaker, i.e. x_1 is a record-breaker by definition.

From a set of simple considerations, it is possible to derive elegant expressions for the number of record-breakers expected in a record of n observations. First, consider the probability of the ith event being a record-breaker. For this purpose, imagine an urn as in Figure 12.2 in which there is initially one black ball only, and that black ball represents a record-breaker.

Now define an experiment as the act of drawing a ball at random from the urn, noting the colour, and returning the ball to the urn together with an additional white ball. The white balls now represent non-record-breaking events. Repeat the experiment n times. While the number of white balls in the urn increases, there is always only one black ball, and therefore the probability of drawing the black ball in the ith experiment is:

$$P(\text{Black in the } i\text{th experiment}) = P(\text{Record-breaker}) = \frac{1}{i} \tag{12.3}$$

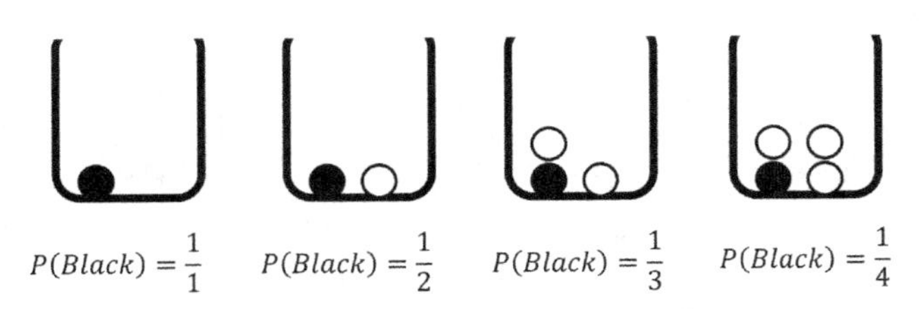

Figure 12.2 Urn experiment where the black ball represents a record-breaker event and the white balls represent non-record-breaker events.

This shows that the longer the record, the lower the probability of observing a record-breaker is, which makes intuitive sense. To count the number of record-breakers in a n year record, a new random variable Y_i is introduced, which indicates if the ith observation is a record-breaker as:

$$Y_i = \begin{cases} 0 & \text{i'th observation IS NOT a record-breaker} \\ 1 & \text{i'th observation IS a record-breaker} \end{cases} \tag{12.4}$$

Using the result that $P(Y_i = 1) = 1/i$ and $P(Y_i = 0) = 1 - 1/i$ the mean value and variance of Y_i can now be calculated with reference to Eqs. 2.17 and 2.19:

$$\begin{aligned} E(Y_i) &= 0 \times P(Y_i = 0) + 1 \times P(Y_i = 1) = \frac{1}{i} \\ V(Y_i) &= \left(0 - \frac{1}{i}\right)^2 \times P(Y_i = 0) + \left(1 - \frac{1}{i}\right)^2 \times P(Y_i = 1) = \frac{1}{i} - \frac{1}{i^2} \end{aligned} \tag{12.5}$$

Finally, it is now possible to derive the mean and variance of the number of record-breakers in a n year record. For this purpose, a new random variable denoted R_n (the subscript n refers to record length) is introduced which simply adds all the values of Y_i as:

$$R_n = \sum_{i=1}^{n} Y_i \tag{12.6}$$

Using the mean and variance operators introduced in section 5.2, noticing that Y_i and Y_j are independent gives:

$$\begin{aligned} E(R_n) &= \sum_{i=1}^{n} E(Y_i) = \sum_{i=1}^{n} \frac{1}{i} \\ V(R_n) &= \sum_{i=1}^{n} V(Y_i) = \sum_{i=1}^{n} \frac{1}{i} - \sum_{i=1}^{n} \frac{1}{i^2} \end{aligned} \tag{12.7}$$

Interestingly, the mean value of R_n is the harmonic series which is divergent in the infinite, i.e. as n converges towards infinity so will R_n, ($E(R_n) \to \infty$ for $n \to \infty$), even if it does so very slowly. This means that, in principle, there *will always* be a new record-breaker sooner or later, even if we might have to wait a very long time. It is therefore probably not wise to be too surprised by the occurrence of a record-breaker. Also, basing engineering design on the largest event on record might not be wise, as an even larger event is likely to occur at some point in time. For example, do not design the height of a flood wall based on the single largest event from the past 10 years as it is guaranteed to be overtopped in the future, even if we cannot say exactly when.

Returning to the example from the River Derwent in Figure 12.1 where a total of eight record-breakers were observed in a $n = 50$ year period. The expected number of record-breakers is:

$$E(R_{50}) = \sum_{i=1}^{50} \frac{1}{i} = 1 + \frac{1}{2} + \frac{1}{3} + \ldots + \frac{1}{50} = 4.499$$
$$V(R_{50}) = \sum_{i=1}^{50} \frac{1}{i} - \sum_{i=1}^{50} \frac{1}{i^2} = 2.8741 \qquad (12.8)$$

The expected number of events is lower than the actually observed number of record-breakers (8), which could have happened by chance, or perhaps a consequence of changes to the conditions generating extreme flood events over time; for example, climate change might have caused an increase in the occurrence of extreme floods in this region.

12.4 A PERFECT STORM

The idea of a perfect storm, where several well-known factors come together to create a uniquely adverse outcome, comes from the book *The Perfect Storm* by Sebastian Junger (1997), which was made into a film by the same name in 2000.

The book and movie tell the story of a storm hitting the north-east Atlantic seaboard of the United States, killing 12 fishermen. The storm was a rare conjuncture of three well-known meteorological phenomena occurring at the same time (a storm starting over the continental United States, a cold front coming from the north, and the tail of a tropical storm). While each of the three phenomena is well-known, the conjuncture of all three at the same time is considered extremely rare and, in this case, resulted in a catastrophic outcome.

From the perspective of probability theory, the conjuncture of several independent events is defined as the union $\cap$ of the events. Consider for example three independent events A, B, and C, and the union of these three events is then given as:

$$P(A \cap B \cap C) = P(A)P(B)P(C) \qquad (12.9)$$

As each of the individual probability terms on the right-hand side of the equation is a number between 0 and 1, the left-hand side of the equation (i.e. the intersection of the three events) will be lower than any of the individual probabilities except for the trivial case of event probabilities being equal to 1 or 0. The concept is shown on Figure 12.3 in the form of a Venn diagram considering the intersection $\cap$ of five independent events $(A, B, C, D, \text{ and } E)$.

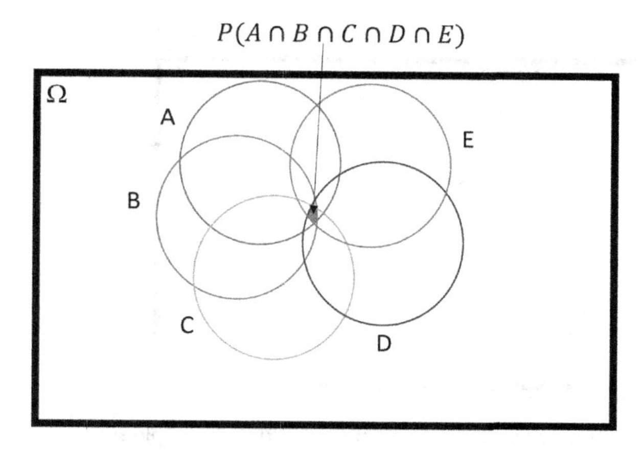

Figure 12.3 Venn diagram showing the perfect storm (area in centre included in all circles).

The sample space is represented by Ω. Again, the area representing the intersection between the events is smaller than the area of any of the individual events, suggesting a lower probability.

12.5 BLACK SWAN EVENTS

The concept of a black swan event was promoted by Nassim Taleb in his book from 2007 titled *The Black Swan: The Impact of the Highly Improbable*. The premise of the book is that traditional statistical methods cannot adequately capture the risk of events that have not previously been observed or anticipated. The name *Black Swan* comes from the idea that before Dutch sailors arrived in Australia in the 17th century, Europeans considered all swans to be white; probably no one ever even considered the possibility that an adult swan could be anything other than a white bird. However, the discovery of a single black swan in Australia resulted in a fundamental shift in this belief; from now on a black swan was no longer an unthinkable event. A black swan event is therefore synonymous with an event we did not even anticipate before it actually happened. The reason for this could be manifold, including careless planning or simply lack of imagination. The terror attacks in New York on 11 September 2001 are often cited as a black swan event.

Traditional probability theory is not well suited to dealing with completely unanticipated events. How can we calculate the probability of something we do not even know the existence of? One possible interpretation is that the first step in every probability calculation is to define the sample space Ω representing all possible events. If an event A is unanticipated it must mean that is it located outside the sample space, i.e. $A \not\subset \Omega$. Once the existence

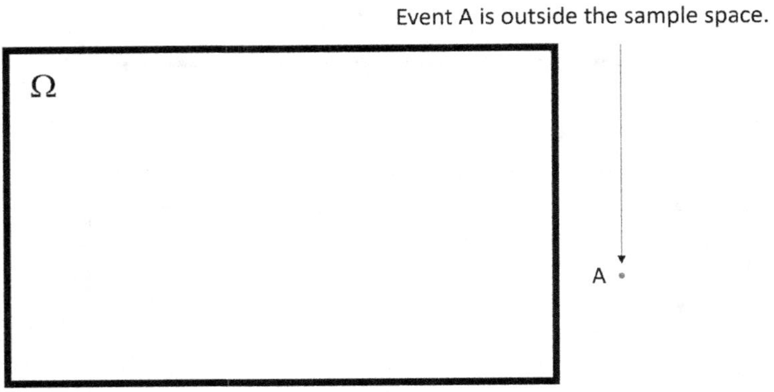

Event A is outside the sample space.

Ω

A

Figure 12.4 The black swan event A located outside the sample space Ω.

of event A has been revealed, then the sample space Ω can be extended to include A in future risk assessments. In hindsight it is perhaps obvious that event A could occur, even if no one anticipated it at the time.

12.6 RISK ASSESSMENT AND MANAGEMENT

There is no universally agreed definition of the term *risk*. According to the Oxford Dictionary risk is defined as "a situation involving exposure to danger." In the technical literature the most common definition of risk is likelihood multiplied by consequence, i.e. the risk of event A is:

$$Risk(A) = Probability(A) \times Consequence(A) \tag{12.10}$$

which is analogous to the definitions of expected payoff or expected utility discussed in Chapter 9. However, sometimes risk is defined as just the likelihood, so it is important to understand the context in which the term is being used.

Most activities in a professional environment require by law a risk assessment to be conducted. A risk assessment is not about creating huge amounts of paperwork, but rather about identifying sensible measures to control the risks in your workplace. This is particularly important in areas where failure to consider hazards can have severe consequences such as most civil engineering site work. Excellent resources on health and safety are available from the UK *Health and Safety Executive* (*http://www.hse.gov.uk/index.htm*).

The steps required for a risk assessment are:

1. Identify undesirable events.
2. Assess probability or likelihood of events.

3. Assess consequences of events.
4. Calculate risk of events (likelihood times consequences).
5. Identify unacceptable risks.
6. Implement mitigation strategies to remove or reduce risks to acceptable levels.

An example of a visual representation of risk assessments is shown in Figure 12.5 as a risk matrix, where each cell contains the product of the likelihood and the consequence. Note that the likelihood is described using qualitative measures such as "not likely," "possible," "likely," and "very likely" and in order of severity with a score from 1 to 4. This is often adequate for the tasks at hand, but it is important to consider good risk communication as covered in section 13 arguing that different people might

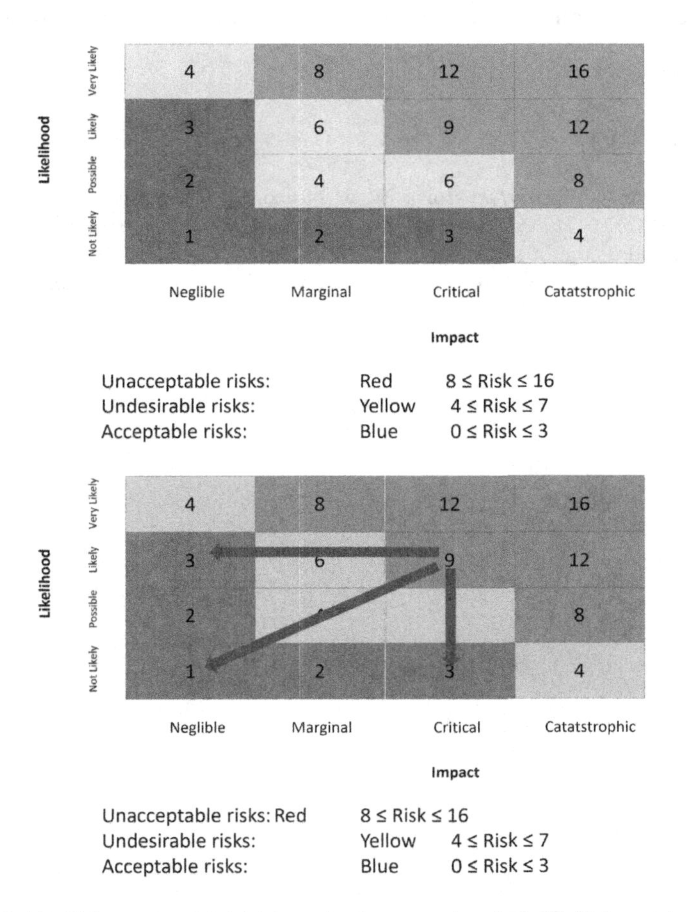

Figure 12.5 (A) Risk matrix with initial scoring between 1 and 16. (B) Risk matrix including possible mitigation routes from unacceptable to acceptable.

assign different levels of probability to qualitative statements. The consequences are similarly measured using a qualitative scale using terms such as "catastrophic," "critical," "marginal," and "negligible." As for likelihood, the consequences are assigned scores from 1 to 4 in order of severity. The final risk score for each cell is therefore a risk number between 1 to 16, with 16 indicating a "very likely and catastrophic" outcome. In this example, a risk score of 8 or above is considered unacceptable, a risk score between 4 and 7 is undesirable, and a risk score below 4 is acceptable.

If events are assigned a score in excess of 4, then it is necessary to enact mitigation strategies to either reduce the consequences, or the likelihood – or ideally both – in order to lower the risk score to acceptable levels. The process of identifying, scoring, and proposing mitigation measures should be captured in a document as an audit trail of health and safety measures taken to avoid preventable accidents. This document should be considered a live document and should be revisited frequently and when circumstances dictate.

REFERENCES

Junger, S. (1997). *The perfect storm*. W. W. Norton & Company.

Taleb, N. (2007). *The black swan; the impact of the highly improbable*. Harlow, England: Penguin Books.

Weaver, W., (1948). Probability, rarity, interest, and surprise. *The Scientific Monthly*, 67(6), pp.390–392.

Chapter 13

Risk Communication

13.1 INTRODUCTION TO RISK COMMUNICATION

Communication is often defined as the act of transferring information from one source (one or more persons, or an organisation) to one or more other persons. Good communication thus is achieved when the persons receiving the information understand it as intended by the communicator. Before discussing communication in more detail, common reasons for communicating risk in civil engineering are reviewed. Firstly, there is communication between risk specialists with a common understanding of the technical language and methods. This type of communication can be through technical reports or design codes, for example. It is relatively straightforward, requiring mostly clarity of presentation. Communication of risk to an audience of lay people or a wider public audience is more challenging and requires an understanding of how people without in-depth training in probability and statistics might perceive and understand complex information involving uncertainties. There can also be different motivations for communicating, including an agenda of persuasion or the desire to provide helpful and important information to the best of our knowledge; we shall assume the latter here and leave the former to others.

There are many examples of bad or unsuccessful communication, where the intended message or information was misunderstood, or only partly understood with a range of consequences to follow. For example, the $125M NASA Mars Climatic Orbiter reportedly failed in September 1999, allegedly because of miscommunication of what scientific units were used in the software controlling the orbiter's thrusters. This is an example of a relatively simple miscalculation that could have easily been avoided by following a better technical protocol. Much more complicated is the communication of risk and uncertainty into the public realm populated by multiple stakeholders with often contrasting levels of education, world views, and social and economic interests. Examples include communication following major infrastructure failures and natural disasters; why did it happen, and who is to blame?

DOI: 10.1201/9781032700373-13

When communicating risk and uncertainty, it is helpful to remind ourselves that the risk of an event A is defined as the product of the likelihood (or probability) $P(A)$ and the consequence $C(A)$ of the event, i.e.:

$$Risk(A)=P(A)C(A) \tag{13.1}$$

Communication of risk therefore requires an understanding of how the recipient of the communication will perceive the conveyed information. This is a complex problem that spans several scientific disciplines, including sociology and psychology. Thus this chapter gives only a brief introduction, and the reader is encouraged to undertake further self-studies into this fascinating topic.

13.2 COMMUNICATING PROBABILITIES

Most technical documents introducing risk assessment, including this book, do so from the perspective of the professional engineer au fait with rational decision-making based on scientific principles. However, from a psychological perspective, most people do not evaluate perceived risk from a purely rational perspective. Research has shown that uncertainty (the inability to predict or control future events) are central features of experiences perceived as stressful.

A technical representation of uncertainty involves a range of possible values or outcomes, for example in the form of a probability density function (pdf) or a confidence interval. While these concepts are familiar to analysts familiar with probability and statistical analysis, the wider public might perceive this differently than initially intended. Johnson and Slovic (1998) highlighted a number of issues that can characterise the response of the wider public to information being presented as a range rather than a single number, for example:

- Perceiving all values in the range as being equally likely.
- Seeing large values (worst case) as being more likely than other options.
- The communicator is incompetent, and therefore unable to provide a correct answer.
- The communicator is not trustworthy and trying to hide the truth or stall decision-making.

A verbal description of probability and uncertainty is also complicated by the fact that different people might perceive such terms differently. On the other hand, a single numerical value might give a wrong impression of certainty on behalf of the communicator. For example, based on work by intelligence officers in NATO in the 1960s concerned about the problem of communicating probabilities (Barclay, 1977), an experiment was conducted where 23 intelligence officers were asked to assign a numerical probability (expressed as a percentage) to each of the statements: "It is highly likely that the Soviets will invade Czechoslovakia" and "It is almost certain that the Soviets will invade Czechoslovakia," and "We believe that the Soviets will invade Czechoslovakia." The message in each message was the same (invasion) but the language varied. The range of answers is summarised in Figure 13.1.

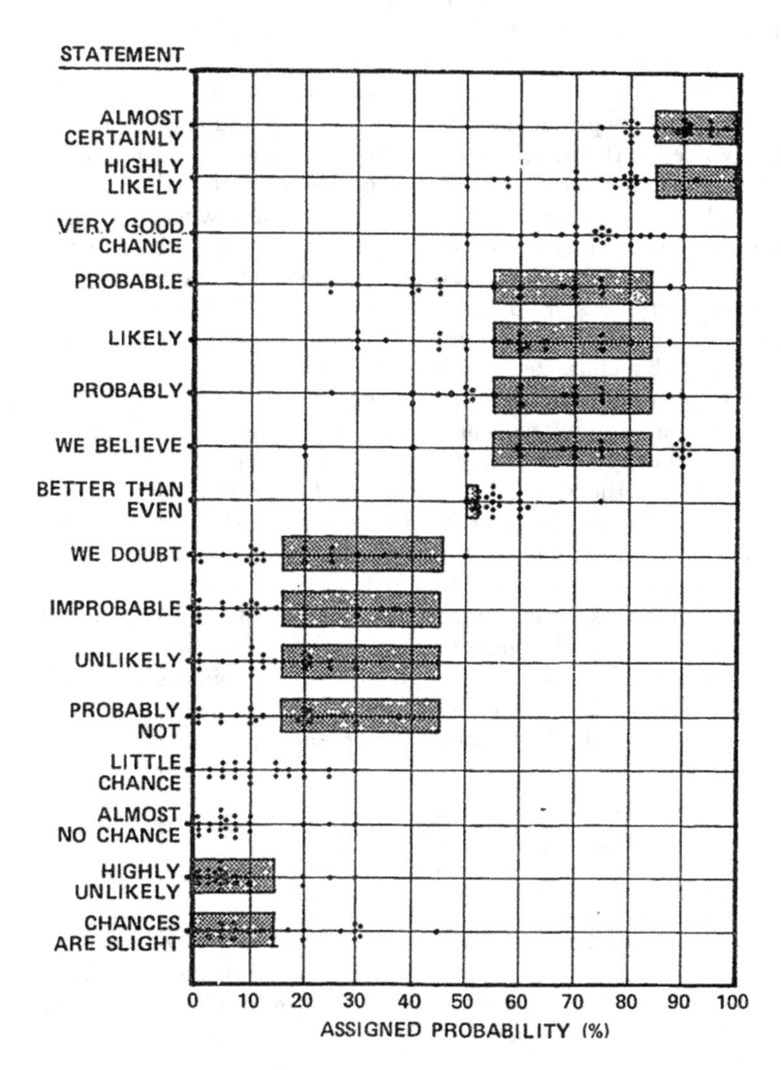

Figure 13.1 Numerical range of probabilities (percentages) assigned to verbal descriptions of probabilities.

Source: Adapted from Barclay (1977).

Following this work, the British intelligence service used a Probability Yardstick (see Figure 13.1) to emphasise that assessments of facts and events are often clouded in uncertainty. The use of language rather than numerical numbers therefore highlights that these assessments are based on a high degree of judgement rather than data analysis. The probability yardstick splits the probability range 0–1 into 7 bins, each with a width defined by a probability range defined as fractions and a verbal description. For example, events that are characterised as "highly unlikely" are deemed to have a probability of occurrence between 0.10 and 0.20 (1:10 to 1:5). This scale is based on research on how most people would place these terms on the numerical probability scale, so might have wider use beyond the intelligence community. The gaps between the bins are fuzzy and emphasise the subjectivity involved in the translation of verbal descriptions to quantitative probabilities (expressed here as either percentages or fractions).

Examples from other areas beyond intelligence where communication of risk and uncertainty is important include, for example, climate change impacts and medical science – both topics that have potentially important aspects on everyone's life. The Intergovernmental Panel on Climate Change (IPCC, 2005) issued guidelines for translating between probabilities and the associated verbal characterisation, as shown in Table 13.1.

Another example of using verbal language to express probabilities and ranges is the attempt by medical agencies to communicate the likely negative side effects of drugs. This is an area where there is propensity for perceiving the worst case as the most likely outcome. The National Institute for Health and Care Excellence (NICE, n.d.) advise the translation between descriptions and probability when communicating the probability of adverse reactions to drugs as shown in Table 13.2.

Finally, when presenting numerical probabilities linked to events, it is also important to consider the underlying message. For example, instead of saying "there is x probability of something happening," it might be equally valid to state that "there is $(1-x)$ probability of it not happening." For example, there is 0.10 probability of a large flood overtopping a defence in any one year, or

Table 13.1 IPCC (2005) Proposed Translation between Verbal Terms and Numerical Probabilities

Terminology	*Likelihood*
Virtually certain	More than 0.99 probability
Very likely	More than 0.90 probability
Likely	More than 0.60 probability
About as likely as not	0.33–066 probability
Unlikely	Less than 0.33 probability
Very unlikely	Less than 0.10 probability
Exceptionally unlikely	Less than 0.01 probability

Table 13.2 Probability of Adverse Reactions to Drugs (According to NICE)

Very Common	*Greater Than 1 in 10*
Common	1 in 100 to 1 in 10
Uncommon	1 in 1000 to 1 in 100
Rare	1 in 10 000 to 1 in 1000
Very rare	Less than 1 in 10 000

there is 0.90 probability of no flood overtopping the wall – the same information but presented in two contrasting ways.

13.3 PUBLIC PERCEPTION OF CONSEQUENCE

When trying to understand how people might perceive risk estimates, it is also important to consider how undesirable, or risky, events are perceived by people. Research by Slovic (1987) argued that these perceptions are influenced by a number of factors, including voluntariness, dread, knowledge, and controllability.

Voluntariness, and to some extent, controllability are connected and represent the degree to which risk exposure is a result of a voluntary action. Slovic (1987) argues that people are likely to accept risks from voluntary activities that are about 1000 times greater than involuntary risks. For example, most people accept the risk of driving a car, but would be loath to live next to a nuclear power plant, even if the risk being involved in a car accident is many times larger than the risk of a nuclear accident.

Dread and knowledge associated with hazardous events are key to understanding public risk perception. An individual might dread a hazard based on perceptions such as lack of control and potentially catastrophic consequences. Lack of knowledge is also a major factor, in particular lack of knowledge about the potential consequences, both the type of consequence as well as the time frame for manifestation; for example, a new or emerging hazard, such as new chemicals and radiation, where the effects are unknown, including the time for manifestation of the risks. Figure 13.3 is adapted from Slovic (1987) and shows the estimated degree of dread and unknown risk for a range of hazards, reflecting the time of publication. The upper right quadrant of the figure includes hazards associated with a high degree of dread and classified as high unknown risk. These hazards include events such as DNA technology (seen as new and unknown in the 1980s) as well as cataclysmic events such as nuclear accidents and nuclear war (again, remember this figure reflects the time at the end of the Cold War). Events with low dread and perceived as having knowable consequences can be found in the lower left quadrant and include more mundane accidents involving items such as chainsaws, bicycles, and trampolines. Few of the hazards are directly

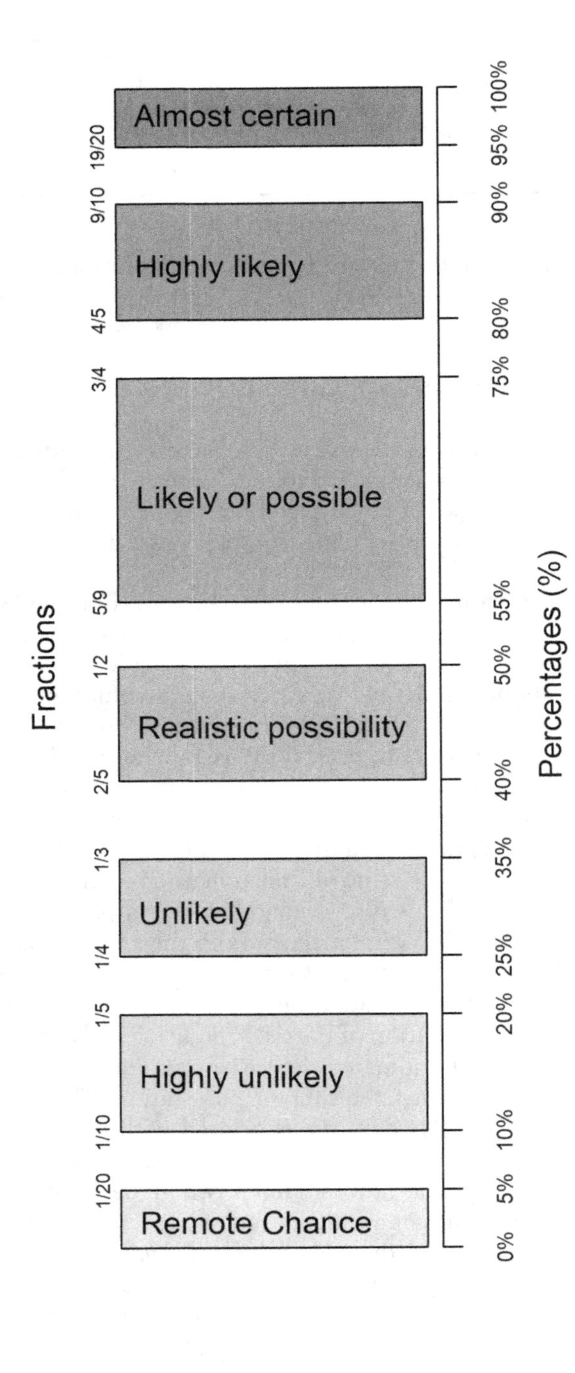

Figure 13.2 The Probability Yardstick.

Source: Adapted from PHIA (2019).

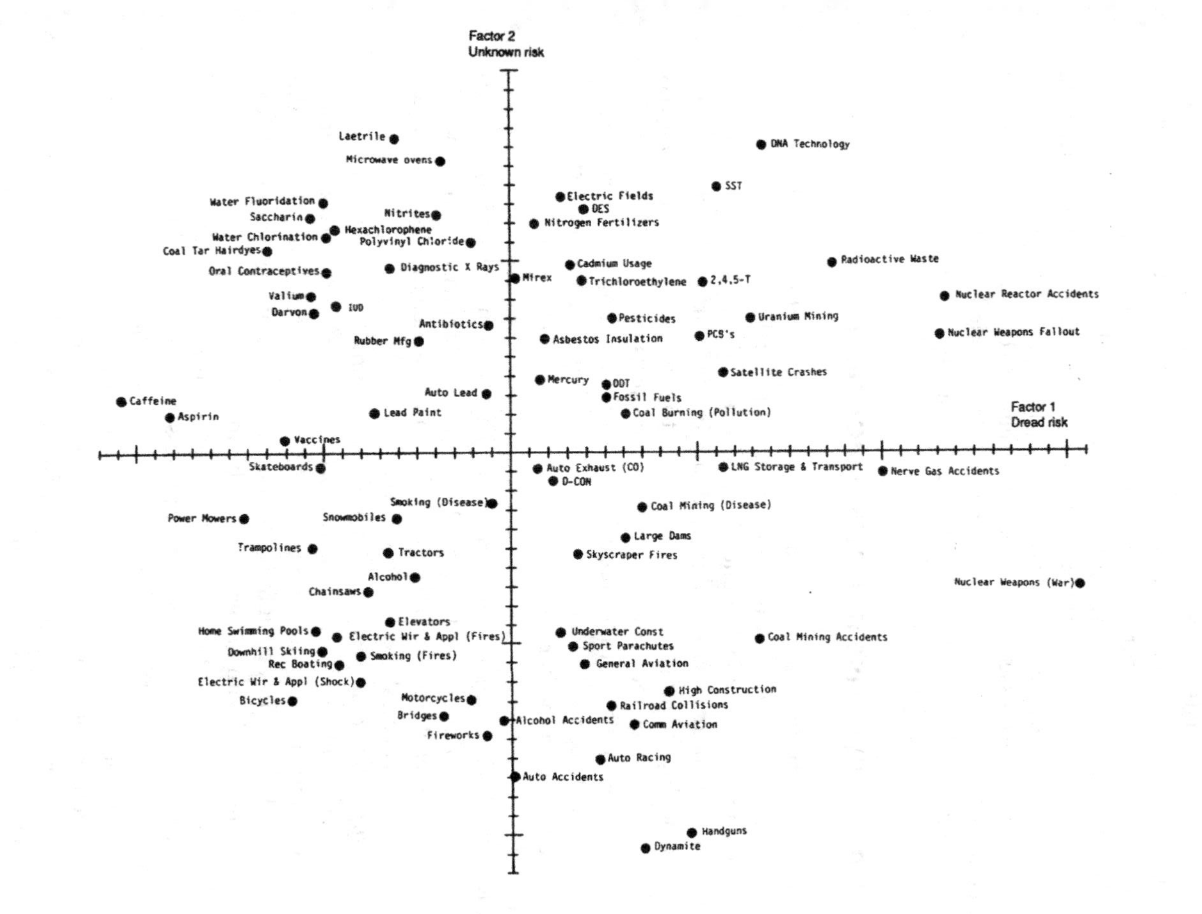

Figure 13.3 Hazards scored according to dread (horizontal axis) and unknown risk (vertical axis).

Source: Adapted from Slovic (1987).

related to civil engineering, but where they appear (bridges, fire in skyscrapers), the consequences are generally considered to be known, and therefore they are not perceived as high risk compared to other emerging risks.

Finally, individuals with strong views about a topic might not necessarily change those views when confronted with new scientific data and information. There can be several reasons for this, including distrust in the organisation or government providing the information. Also, initial strong views might influence how new data are perceived, such that new data contrasting these views will be perceived with scepticism, while data that contradict these views are dismissed or considered unreliable. This phenomenon is also sometimes known as confirmation bias, where people look for evidence to support existing beliefs while dismissing contrasting evidence.

13.4 BEST PRACTISE IN COMMUNICATION

Good communication is vitally important in all parts of modern life in both private and public realms. Consequently, it is a topic that is covered in multiple academic and practical publications with a plethora of advice to follow. When communicating risk and uncertainty in the context of civil engineering specifically, the major challenge is often to inform the public about risk and hazards associated with infrastructure projects such as safety of buildings and exposure to natural hazards such as floods and landslides. In summary, the most common features of good risk communication consist of the points below in some iteration:

Know what you are trying to achieve: Why are you embarking on communication in the first place? Do you want to inform the audience, or are you reaching out to seek views? Can you articulate how you would measure successful communication?

Use appropriate language: Assuming the communication is aimed at a larger public audience, it is important to use appropriate language, avoiding jargon and complex technical details. However, this is a difficult balance to get right, providing clear messages without omitting details, which can be perceived as hiding facts.

Be honest about uncertainties: It is important that uncertainties are made clear to the audience as part of professional and ethical best practice. However, uncertainty also implies unknown outcomes, which is not always welcome. Firstly, increased unknown outcomes can increase the perception of something as involving more risk or having higher consequences. Secondly, it can be difficult to understand that is understood by uncertainty, or many people will assume the worst outcome is the most likely. Other issues such as credibility, competence, and hidden agendas of the communicator can emerge. If possible, it can be helpful to support messages of risk and uncertainty with visual graphics to help convey messages.

Be transparent: If information is not shared by the body managing the risk, it is likely the public will obtain information from other sources. For example, it might be useful to acknowledge gaps in knowledge and highlight ongoing efforts to gather more data or refine risk assessments. This is also an opportunity to ask for feedback, thus increasing inclusiveness.

Focus on mitigation and resilience: Highlight how communities and individuals can take action to mitigate or lessen the negative impacts of undesirable events. Remember that a feeling of being in control can greatly increase the degree of acceptance.

Remember listening as well as talking: Listening to people and adopting an interactive style of communication will aid in bridging the gap between public and expert perceptions of risk. Whether scientifically accurate or not, the public's perception is reality; if the public believe a risk exists, it can be expected they will act in accordance with that belief. Monitoring public opinion about a risk is also key to treating the public as a partner.

REFERENCES

Barclay, S. (1977). *Handbook for decision analysis. Technical report TR-77-6-30*, Office for ~Naval Research, Washington, DC.

Intergovernmental Panel on Climate Change (2005). *Guidance notes for lead authors of the IPCC fourth assessment report on addressing uncertainties*. Intergovernmental panel on Climate Change.

Johnson, B. B. and Slovic, P. (1998). Lay views on uncertainty in environmental health risk assessment. *Journal of Risk Research*, 1(4), pp. 261–279.

NICE (n.d.). https://bnf.nice.org.uk/medicines-guidance/adverse-reactions-to-drugs/

Professional Head of Intelligence Assessment (2019). Professional development framework for all-source intelligence assessment. *First Edition*. https://assets.publishing.service.gov.uk/media/6421b6a43d885d000fdadb70/2019-01_PHIA_PDF_First_Edition_Electronic_Distribution_v1.1__1_.pdf

Slovic, P. (1987). Perception of risk. *Science*, 236(4799), pp. 280–285.

Chapter 14

Useful EXCEL Functions

14.1 STATISTICAL DISTRIBUTIONS

The EXCEL functions introduced in this section all calculate the probability density function (pdf), the probability mass function (pmf), or the cumulative distribution function (cdf) of selected statistical distributions used throughout this textbook and commonly encountered in civil engineering risk analysis. All functions that evaluate pdf and cdf of a distribution use the format exemplified in Figure 14.1, where the "Cumulative" input box has two options:

Cumulative = TRUE: return cdf
Cumulative = FALSE: return pdf

14.1.1 The Uniform Distribution and the RAND Function

The uniform distribution has a pdf defined on an interval $[a;b]$ as:

$$f(x) = \frac{1}{b-a} \tag{14.1}$$

and zero elsewhere. The corresponding cdf is:

$$F(x) = \frac{x-a}{b-a} \tag{14.2}$$

The function *RAND* will generate a random number from a uniform distribution defined on the interval $[0;1]$ and requires no input. So just type "=*RAND()*" into a cell and watch EXCEL generate a random number between 0 and 1.

To generate a random number from a uniform distribution defined more generally on the interval $[a;b]$, the following transformation of the *RAND* function can be used:

DOI: 10.1201/9781032700373-14

$$a + (b - a) \times RAND(\ ) \tag{14.3}$$

RAND is a volatile function that will change value every time the spreadsheet is updated. The following is a tip from the EXCEL help page: "*If you want to use RAND to generate a random number but don't want the numbers to change every time the cell is calculated, you can enter =RAND() in the formula bar, and then press F9 to change the formula to a random number. The formula will calculate and leave you with just a value.*"

14.1.2 The Normal Distribution

The two-parameter normal distribution has a pdf defined as:

$$f(x) = \frac{1}{\sqrt{2\pi}} \frac{1}{\sigma} exp\left(-\frac{1}{2}\left(\frac{x-\mu}{\sigma}\right)^2\right) \tag{14.4}$$

where the μ is the mean and σ is the standard deviation (*Standard_dev*). The corresponding cdf does not have an explicit analytical expression.

A special case of the normal distribution is the standardised normal distribution ($\mu = 0$ and $\sigma = 1$) which has a pdf defined in Chapter 2 with a Greek letter φ (phi) and a mathematical expression:

$$f(z) = \frac{1}{\sqrt{2\pi}} exp\left(-\frac{1}{2} z^2\right) \tag{14.5}$$

Again, the corresponding cdf (Φ) has no explicit analytical expression but is often defined as:

$$F(z) = \Phi(z) \tag{14.6}$$

For example, consider a random variable $X \sim N(0,1)$ for which $P(X \leq 1.2816) = 0.9000$ according to standard statistical tables such as Appendix A of Chapter 2.

EXCEL has a number of built-in functions supporting the calculations involving the normal distribution. A summary is shown in Table 14.1, and screenshots showing the user interface for each function are shown in Figures 14.1–14.5.

14.1.3 The Log-Normal Distribution

The pdf of the log-normal distribution is defined as:

$$f(x) = \frac{1}{x\beta\sqrt{2\pi}} exp\left(-\frac{1}{2}\left(\frac{\ln(x) - \alpha}{\beta}\right)^2\right) \tag{14.7}$$

Table 14.1 Summary of EXCEL Functions

EXCEL Function	*Description*	*Figure Reference*
NORMDIST	Pdf and cdf of normal distribution	14.1
NORM.INV	Quantile of normal distribution	14.2
NORM.S.DIST	Pdf and cdf of standard normal distribution	14.3
NORM.S.INV	Quantile of standard normal distribution	14.4
GAUSS	Return probability of a member falling between the mean and X standard deviations from the mean	14.5

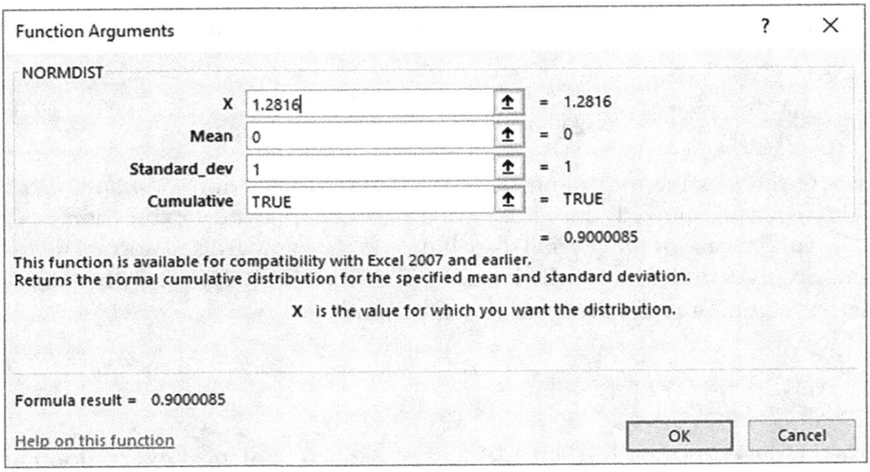

Function Arguments ? X

NORMDIST

X	1.2816	= 1.2816
Mean	0	= 0
Standard_dev	1	= 1
Cumulative	TRUE	= TRUE

= 0.9000085

This function is available for compatibility with Excel 2007 and earlier.
Returns the normal cumulative distribution for the specified mean and standard deviation.

X is the value for which you want the distribution.

Formula result = 0.9000085

Help on this function OK Cancel

Figure 14.1 NORMDIST: Evaluates the normal distribution at a point *X*.

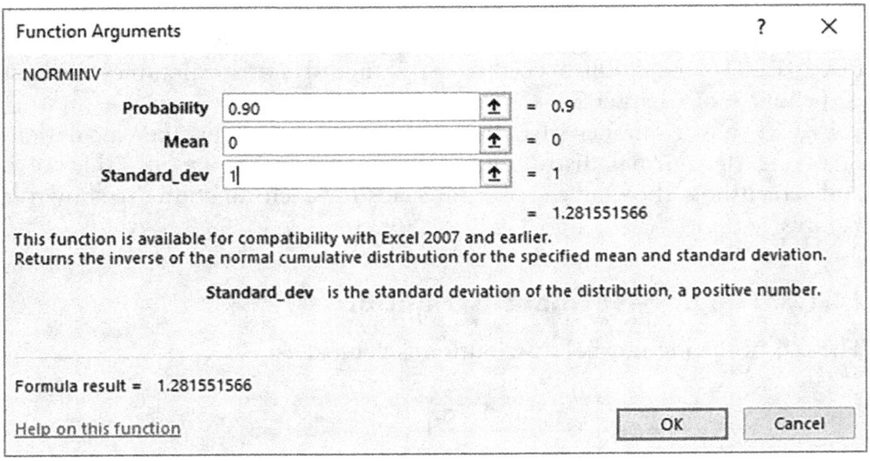

Function Arguments ? X

NORMINV

Probability	0.90	= 0.9
Mean	0	= 0
Standard_dev	1	= 1

= 1.281551566

This function is available for compatibility with Excel 2007 and earlier.
Returns the inverse of the normal cumulative distribution for the specified mean and standard deviation.

Standard_dev is the standard deviation of the distribution, a positive number.

Formula result = 1.281551566

Help on this function OK Cancel

Figure 14.2 NORM.INV: Returns the *p*th quantile of the normal distribution.

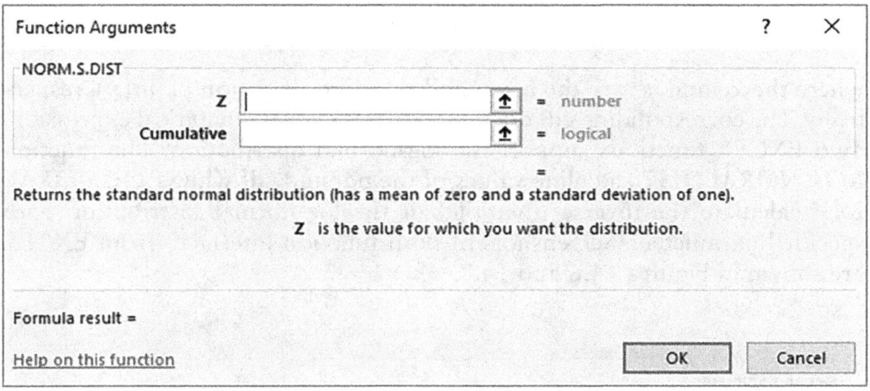

Function Arguments ? ×

NORM.S.DIST

Z = number

Cumulative = logical

=

Returns the standard normal distribution (has a mean of zero and a standard deviation of one).

Z is the value for which you want the distribution.

Formula result =

Help on this function OK Cancel

Figure 14.3 *NORM.S.DIST*: Returns the standard normal distribution at *Z* (has a mean of 0 and a standard deviation of 1).

Function Arguments ? ×

NORM.S.INV

Probability 0.90 = 0.9

= 1.281551566

Returns the inverse of the standard normal cumulative distribution (has a mean of zero and a standard deviation of one).

Probability is a probability corresponding to the normal distribution, a number between 0 and 1 inclusive.

Formula result = 1.281551566

Help on this function OK Cancel

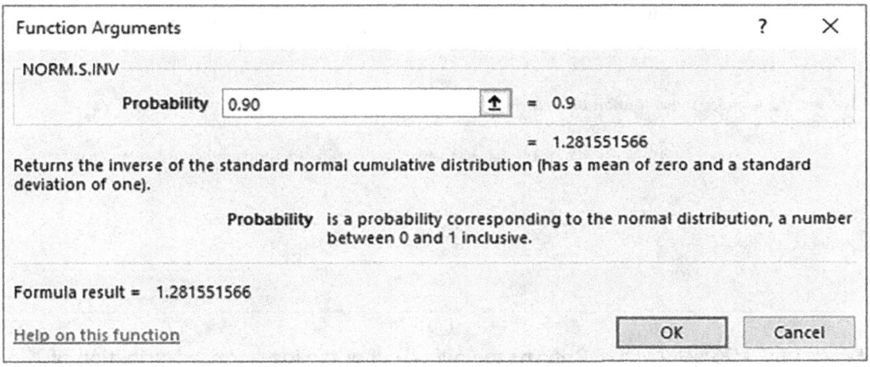

Figure 14.4 *NORM.S.INV*: Returns the inverse (or quantile) of the standard normal cumulative distribution at the point *Z*. The distribution has a mean of 0 and a standard deviation of 1.

Function Arguments ? ×

GAUSS

X 1.96 = 1.96

= 0.475002105

Returns 0.5 less than the standard normal cumulative distribution.

X is the value for which you want the distribution.

Formula result = 0.475002105

Help on this function OK Cancel

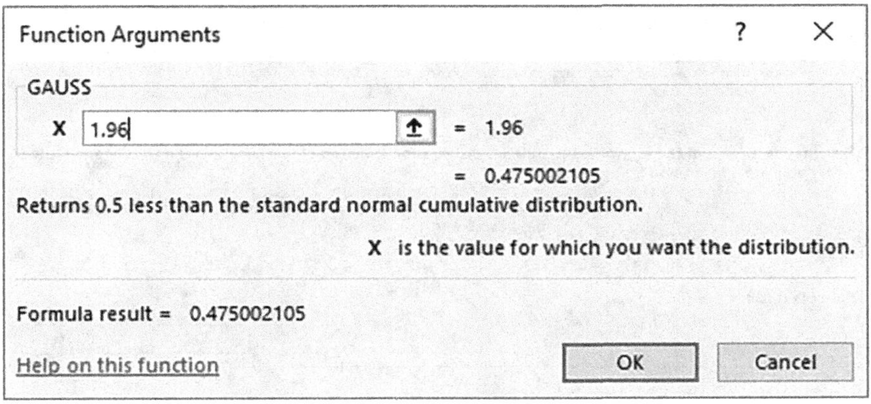

Figure 14.5 *GAUSS*: Calculates the probability that a member of a standard normal population will fall between the mean and z standard deviations from the mean.

where the α and β are the mean and standard deviation of $\ln(x)$. respectively. The corresponding cdf does not have an explicit analytical expression. Two EXCEL functions support the log-normal distribution. The function *LOGNORM.DIST* calculate values of the pdf and cdf while *LOGNORM.INV* calculate the inverse (quantile) of the log-normal distribution with specified parameters. Screenshots of both function interfaces from EXCEL are shown in Figures 14.6 and 14.7.

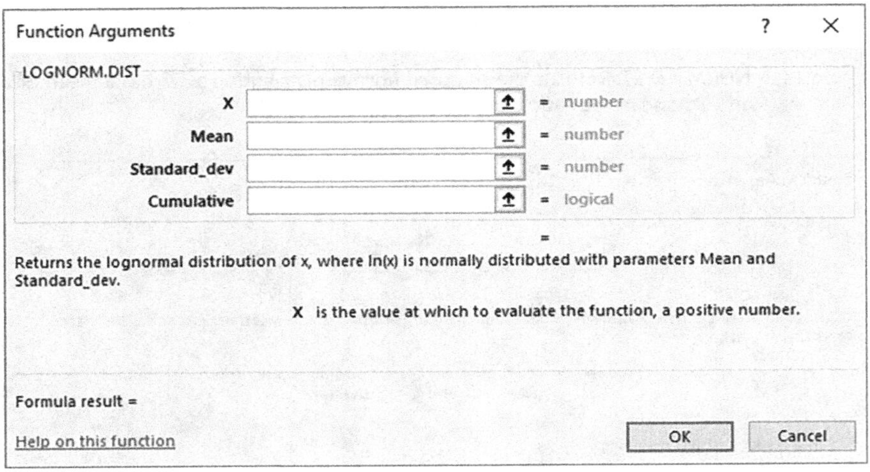

Function Arguments ? ×

LOGNORM.DIST

X = number

Mean = number

Standard_dev = number

Cumulative = logical

=

Returns the lognormal distribution of x, where ln(x) is normally distributed with parameters Mean and Standard_dev.

X is the value at which to evaluate the function, a positive number.

Formula result =

Help on this function OK Cancel

Figure 14.6 LOGNORM.DIST: Returns the pdf or cdf of the log-normal distribution of *X*.

Function Arguments ? ×

LOGNORM.INV

Probability = number

Mean = number

Standard_dev = number

=

Returns the inverse of the lognormal cumulative distribution function of x, where ln(x) is normally distributed with parameters Mean and Standard_dev.

Probability is a probability associated with the lognormal distribution, a number between 0 and 1, inclusive.

Formula result =

Help on this function OK Cancel

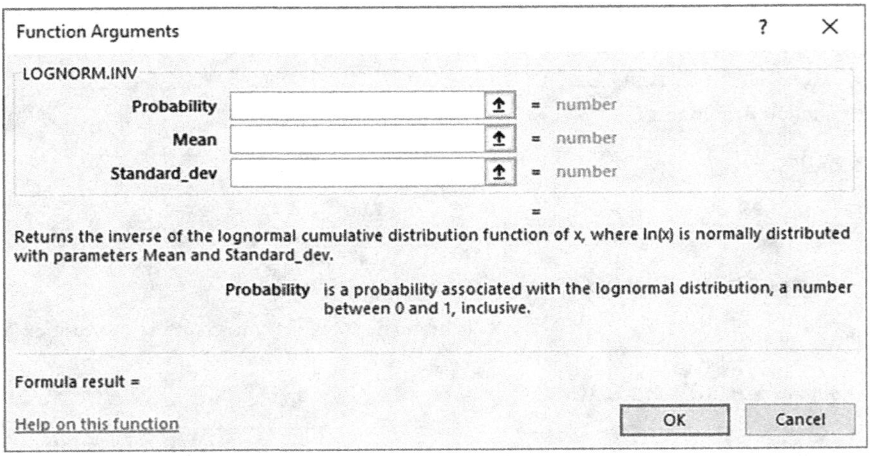

Figure 14.7 LOGNORM.INV: Returns the inverse of the log-normal cumulative distribution function of *X*.

14.1.4 The Gamma Distribution

The pdf of the gamma distribution is defined as:

$$f(x) = \frac{1}{\beta^{\alpha}\Gamma(\alpha)} x^{\alpha-1} exp\left(-\frac{x}{\beta}\right) \tag{14.8}$$

For $\alpha = 1$ the gamma distribution becomes the one-parameter exponential distribution:

$$f(x) = \lambda exp(-\lambda x) \tag{14.9}$$

where $\lambda = \beta^{-1}$.

Three EXCEL functions support the gamma distribution. The function *GAMMA.DIST* calculate values of the pdf and cdf while *GAMMA.INV* calculate the inverse (quantile) of the gamma distribution with specified parameters. Finally, the *EXPON.DIST* function is a special case of the *GAMMA.DIST* evaluating the pdf and cdf of the exponential distribution (as per the definition in Eq. 14.9). Screenshots of the three function interfaces from EXCEL are shown in Figures 14.8 and 14.10.

There is no function in EXCEL for explicitly calculating the inverse of the exponential distribution, but *GAMMA.INV* can be used by setting $\alpha = 1$ and $\beta = \lambda^{-1}$.

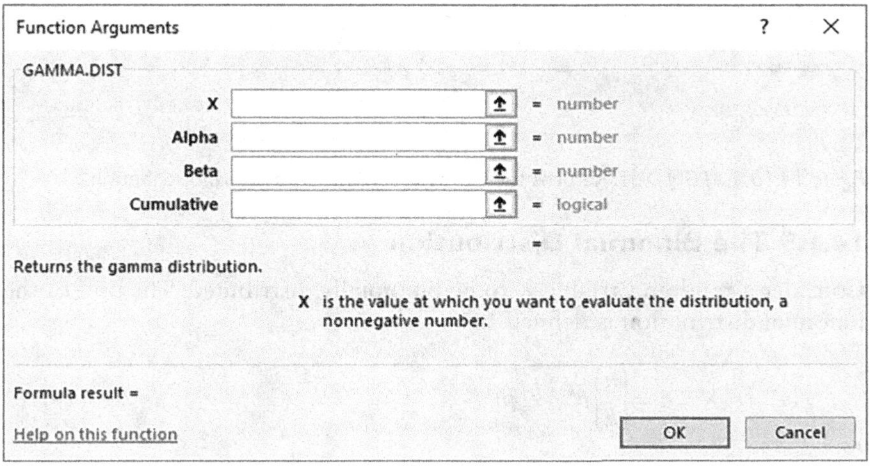

Function Arguments ? ×
GAMMA.DIST
X = number
Alpha = number
Beta = number
Cumulative = logical
=
Returns the gamma distribution.
X is the value at which you want to evaluate the distribution, a nonnegative number.
Formula result =
Help on this function OK Cancel

Figure 14.8 GAMMA.DIST: Returns the pdf or cdf of the gamma distribution.

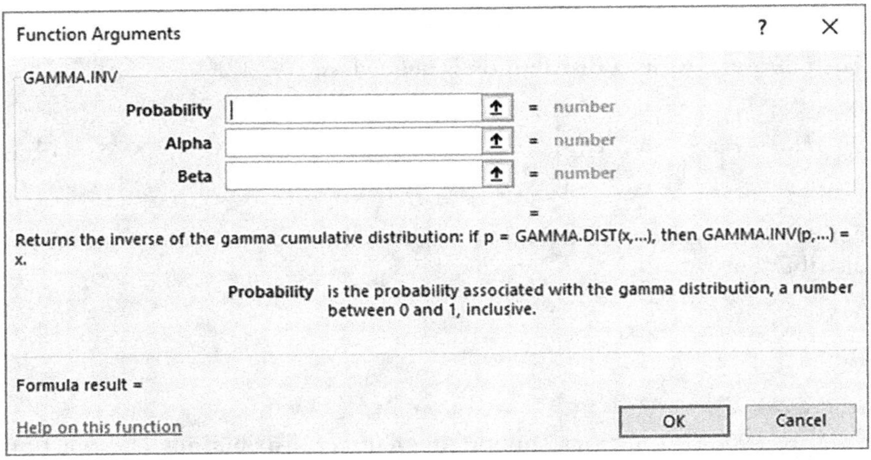

Function Arguments ? ×

GAMMA.INV

Probability = number

Alpha = number

Beta = number

=

Returns the inverse of the gamma cumulative distribution: if p = GAMMA.DIST(x,...), then GAMMA.INV(p,...) = x.

Probability is the probability associated with the gamma distribution, a number between 0 and 1, inclusive.

Formula result =

Help on this function OK Cancel

Figure 14.9 GAMMA.INV: Returns the pth quantile of the gamma distribution.

Function Arguments ? ×

EXPON.DIST

X = number

Lambda = number

Cumulative = logical

=

Returns the exponential distribution.

X is the value of the function, a nonnegative number.

Formula result =

Help on this function OK Cancel

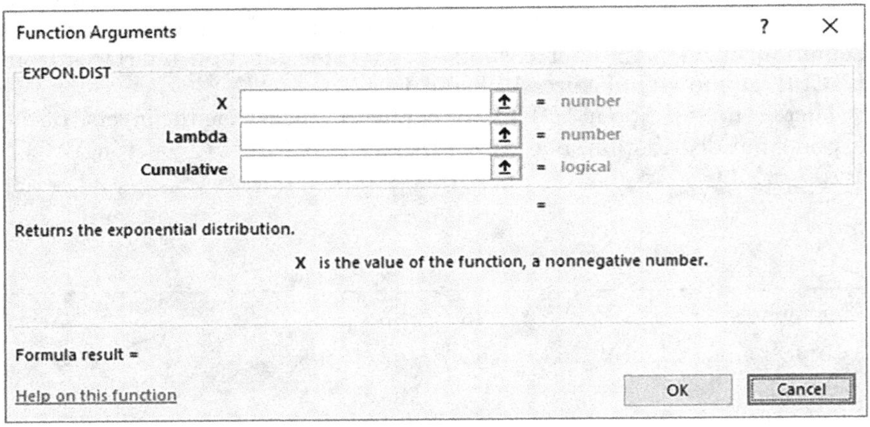

Figure 14.10 EXPON.DIST: Returns the pdf or cdf of the exponential distribution.

14.1.5 The Binomial Distribution

Consider a random variable X to be binomially distributed. The pmf of the binomial distribution is defined as:

$$f(k)=P(X=k)=\binom{n}{k}p^k(1-p)^{n-k} \tag{14.10}$$

which is the probability of selecting k successes from n experiments, each experiment assigned a probability of success equal to p.

The cdf of the binomial distribution is defined as:

$$F(k)=P(X\leq k)=\sum_{i=0}^{k}f(i) \tag{14.11}$$

Printed in the United States
by Baker & Taylor Publisher Services